EOC Algebra 1 Study Guide

"Don't let math ruin your life anymore"
Call us today!
www.ihatemathgroup.com

Have a μath day!

The EOC or End of Course for Algebra 1 is a test given by the Florida Department of Education by the Florida Standards Assessments.

Please visit the Department of Education of Florida for more information about the EOC Algebra 1 test.

http://www.fldoe.org/

I Hate Math Group is not endorsed or affiliated with the Florida Department of Education

I Hate Math Group ®
www.ihatemathgroup.com

"I Hate Math Group®" is a tutoring and test preparation company.

We offer local and online tutoring.

Visit our website for more information.

"Don't let math ruin your life anymore"

#ihatemathgroup

ihatemathgroup.com

All rights reserved. No part of this publication may be reproduced, distributed, or transmitted in any form or by any means, including photocopying, recording, or other electronic or mechanical methods, without the prior written permission of the publisher, except in the case of brief quotations embodied in critical reviews and certain other noncommercial uses permitted by copyright law.
This book is not endorsed nor sponsored by the Department of Education or another institution.

Table of Contents

Real Numbers **10**
    Number Line 11
    Adding and Subtracting Numbers 12
    Multiplying and Dividing Numbers 13
    Absolute Value 14
    Order of Operations 15

Factors and Multiples **17**
    Factors 17
    Divisibility Rules 18
    Multiples 21
    Greatest Common Factor and Least Common Multiple 22

Fractions **26**
    Basic Fractions 26
    Adding and Subtracting 28
    Multiplying and Dividing 31
    Percentages 33
    Increase and Decrease 38
    Percentage Word Problems 40
    Ratios 43

Rates and Proportions **47**

Sets and the Venn Diagram **50**

Probability **55**
    Probability of Independent/Dependent Events 57
    Probability of Mutually Exclusive Events 60
    Probability of Not Mutually Exclusive Events 62
    Factorial Notation 66

    Counting Methods..67
    Combination and Permutation...67

Fundamentals of Statistics...**72**
    Central Measurements..75
    Five-Number Summary..77
    Measurements of Dispersion..80
    Scatter Plots and Correlation...84
  QUICK ARITHMETIC SUMMARY...89

Expressions **93**
    Evaluating Expressions 93
    Simplifying Expressions 98
    Translating Expressions 101
    Exponential Expressions 104
    Multiplying Expressions... 109

Factoring **111**
    Greatest Common Factor 111
    By Grouping.. 112
    Trinomials... 115
    Difference of Squares 118
    Sum and Difference of Cubes.. 121

Rational Expression **123**
    Adding and Subtracting.. 123
    Multiplying and Dividing... 126
    Complex Fractions.. 123

Equations **131**
    Linear Equations 131
    Linear Equations with Fractions.. 132
    Literal Equations 136

Inequalities **140**
- Basic Inequalities 140
- Compound Inequalities 146
- Absolute Value Equations 149
- Absolute Value Inequalities 152

Coordinate Geometry **155**
- Quadrants 156
- Distance and Midpoint 157
- Slope of the Line 159
- Intercepts 165
- Building the Equation of a Line 168
- Graphing a Line 173
- Parallel Lines 181
- Perpendicular Lines 182
- Building an Equation of Parallel and Perpendicular Lines 185

System of Equations **191**
- Solving by Graphing 191
- Solving by Substitution 197
- Solving by Elimination 201

Graphing Inequalities **208**

Radical Expressions **212**
- Basic Operations 212
- Rationalizing Radicals 214
- Radical Expressions with Variables 216
- Radical Equations 218

Quadratic Equations **221**

Quadratic Formula **224**

Complex Numbers..**228**

    Operations of Complex Numbers............................231

    Complex Solutions..233

Polynomials...**235**

Functions..**236**

    Symbolism..239

    Composite Functions..241

Graphing the Quadratic Function............................**243**

    Transformation Rules..249

Symmetry of functions..251

Exponential Functions..**257**

Logarithmic Functions..**259**

    Properties of Logarithms......................................261

Sequences..**264**

    Geometric Sequences..266

Series..**268**

    Sigma Notation...271

    Convergent and Divergent....................................273

Interest..**275**

    Compounding Interest..275

The Circle..**278**

Exponential function Graph....................................**282**

Word Problems..**287**

    Motion Problems...287

    Rate/Work Problems...290

Consecutive Integers..291
Mixture Problems...293
Simple Interest Problems...294
Profit Problems..295
Money Problems...296
Age Problems..298

Practice tests..313
Test 1..313
Answer for Test 1..317
Test 2..320
Answer for Test 2..325
Test 3..328
Answer for Test 3..334
Test 4..339
Answer for Test 4..345
Test 5..353
Answer for Test 5..360
Cheat sheets..368

Real Numbers

Real numbers are subdivided into **integers, rational and irrationals.**

Integers are whole numbers that can be negative or positive.
- **Whole numbers:** -2,-1, 0,1,2,3....
- **Even integers:** -4,-2, **0,** 4, 6, 8, 10
  (0 is an even integer)
- **Odd integers:** -3,-1, 1, 3, 5

Rational: All integers and fractions: 0.5, $\frac{1}{4}$, $\frac{3}{4}$, etc.

Irrational: These numbers have infinite decimals: π, √2, √3

THE NUMBER ZERO "0" IS AN EVEN INTEGER!

In the following sentences classify if True or False

1) Is - 4 an integer?   T/F
2) Is "zero" an integer?   T/F
3) Is "zero" an odd integer?   T/F
4) When you add an odd integer with an even integer, your result will ALWAYS be an odd integer.   T/F
5) When you multiply and odd or an even integer by 2 then your result will ALWAYS be an even integer.   T/F
6) When you multiply two odd integers your answer will ALWAYS be an odd integer. T/F

Answers: 1) T, 2) T, 3) F (It is even) 4) T 5) T 6) T

Number Line

The number "zero" divides the number line. The right side has positive numbers and the left side has negative numbers.

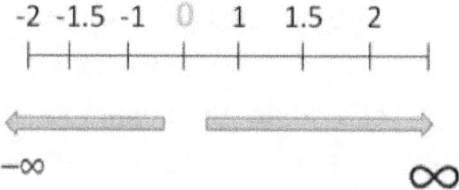

The closer a negative number is to zero, the greater its value.

Examples: -10 is greater than -20
            -2 is greater than -3

Let's practice!
1) 0.980 greater than 0.987   T/F
2) -3 greater than -4?   T/F
3) 0.5 is greater than 0.05   T/F
4) -0.12 is less than -0.13   T/F
5) 0.01 is greater than 0   T/F
6) -120 is less than -100   T/F

Answers: 1) F 2) T 3) T 4) F 5) T 6) T

Adding and Subtracting

When you add two numbers **with different signs, you subtract** and keep the sign of the largest number.

Examples: - 10 + 4 = - 6
- 6 + 4 = - 2

When you add two numbers with the **same sign, you add** and keep the same sign.

Examples: -3 – 2 = - 5
-6 – 9 = - 15

Let's do another example:
-2 + 3 + 4 – 9 – 10 + 6 + 11 =?

Step One: Group the positive numbers and the negative numbers together: **-2-9-10**+3+4+6+11=?

Step Two: Do the operation for the negative numbers:
(-2-9-10 = -21)

Step Three: Do the operation for the positive numbers:
(3+4+6+11= 24)

Final Step: Now group the two numbers together
-21+24 = **3**

Let's practice!
1) 3 + 5 + 10 – 12 – 15 =
2) -6 – 11 + 6 – 12 =
3) -5 – 6 – 8 =
4) 13 – 5 – 6 – 10 =
5) -18 + 23 – 11 =
6) -7 – 7 – 7 + 10 =

Answers: 1)-9, 2) -23, 3)-19, 4)-8, 5)-6, 6)-11

Multiplying and Dividing

When you have to multiply or divide with different signs remember these rules:

$$- \text{ times } - = +$$

$$- \text{ times } + = -$$

Examples:

(-3) times (-2) = 6

(-3) times (2) = - 6

(-6) divided by (-2) = 3

(-6) divided by (2) = -3

Let's Practice!

1) (-1)(1)(-1)(-1)(1) =

2) (-3)(-5)(9)(-1) =

3) (-3)(-4)(-4) =

4) 3(4)(4)(0) =

5) -10/-2 =

6) 10/-2 =

7) 100/-2 =

8) -20/5 =

9) -10(0) =

Answers: 1)-1, 2) -135, 3) -48, 4) 0, 5) 5, 6) -5, 7) -50, 8) -4, 9) 0

Absolute Value

The absolute value of a number is always positive; whatever is inside the absolute value will be positive.

Example: $|-4| = 4$

> WHEN THE NEGATIVE SIGN **IS OUTSIDE THE ABSOLUTE VALUE, THE ANSWER WILL INCLUDE THE NEGATIVE SIGN**

$-|4| = -4$

In this example, the negative is **outside** the absolute value. Therefore, when you find the absolute value of 4 you get positive 4. Remember, the negative is **still there** so the answer is -4.

Let's Practice:

1) $|-4+5-1| =$
2) $|-4-4| =$
3) $-4|-6| =$
4) $4|6| =$
5) $-|3-1|-|5| =$
6) $|-5-10|-4 =$

Answers: 1) 0, 2) 8, 3) -24, 4) 24, 5) -7, 6) 11

Order of Operations

Order of operations determines the steps to be taken to solve an expression.

P.E.M.D.A.S: This mnemonic is used to remember the order of operations.

Order of Operations: You always need to perform the calculations in the following order

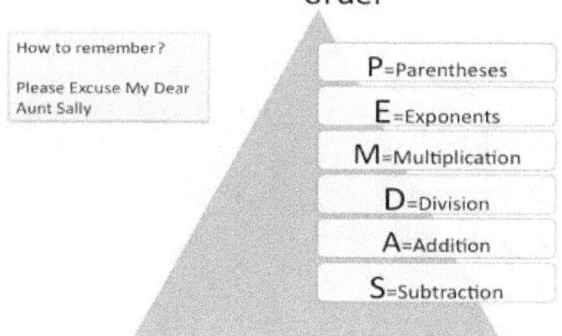

How can you remember P.E.M.D.A.S?
Please **E**xcuse **M**y **D**ear **A**unt **S**ally

Example:
Simplify the following expression:

$(4+2) \div 2 + 3 \cdot 2^2 - 4 + 10^2$

Step 1: **P**arenthesis

$6 \div 2 + 3 \cdot 2^2 - 4 + 10^2$

Step 2: **E**xponents:

$6 \div 2 + 3 \cdot 4 - 4 + 100$

Step 3: **M**ultiplication:

$6 \div 2 + 12 - 4 + 100$

Step 4: **D**ivision:
$$3 + 12 - 4 + 100$$
Step 5: **A**ddition:
$$115 - 4$$
Step 6: **S**ubtraction:
$$111$$

Final Answer: 111

Let's Practice!

1) $10 + (5 - 1)^2 \div 2 \times 7 =$

2) $\dfrac{8 - 1^2}{21} =$

3) $12 \div 3 \times 5 + 7 =$

4) $7(9-3)^2 \div 6 \times (-1) =$

Answers: 1) 66, 2) $\dfrac{1}{3}$ 3) 27, 4) -42

Factors of a number

The factors of a number are the positive integers that evenly divide into that number.
For example, the factors of 6 are 1, 2, 3 and 6.

Let's do a couple of examples:
The factors of 10 are 1,2,5 and 10.
The factors of 30 are 1,2,3,5,6,10,15 and 30.

Factors of a number?

The positive integers that evenly divide into that number

The factors of 12= 1,2,3,4,6,12

Divisibility Rules

Divisibility rules will help you identify the factors of a number.
Let's go over the basic rules.

Divisibility by 2: The last digit is even

Divisibility by 3: The sum of its digits is divisible by 3

Divisibility by 4: The last two digits are divisible by 4

Divisibility by 5: The last digit is either 5 or 0

Divisibility by 6: When a number is divisible by 2 AND 3, then it's divisible by 6

Divisibility by 9: The sum of the digits is divisible by 9

Let's do examples how to find the divisibility of different numbers:

The number 312:

* It is divisible by 2 because the last digit (2) is even.

*It is divisible by 3 because when you add its digits 3+1+2 =6 the total is divisible by 3.

*It is divisible by 4 because its last digits (12) is divisible by 4.
*It is divisible by 6 because it is divisible by 2 and 3.

The number 111:

*It is divisible by 3 because when you add its digits 1+1+1 =3 the total is divisible by 3.

The number 750:

* It is divisible by 2 because the last digit is even

*It is divisible by 3 because when you add its digits 7+5+0 =12 the total is divisible by 3.

*It is divisible by 5 or 0.
*It is divisible by 6 because it is divisible by 2 and 3.

If they ask you **what are the factors of 24**, then you have to start with the question. Can I divide 24 by 1? 2?, 3? and so on until you get to 24.

Let's do it.

- Can I divide it by 1?

Yes, then 1 is a factor.

- Can I divide it by 2?

Yes, then 2 is a factor.

- Can I divide it by 3?

Yes, then 3 is a factor

- Can I divide it by 4?

Yes, then 4 is a factor

- Can I divide it by 5?

No, then 5 is NOT a factor.

Factors of 24: 1,2,3,4,6,8,12,24
Let's Practice:
1) Find the factors of 36:
2) Find the factors of 45:
3) Find the factors of 150:
4) Find the factors of 17:

Let's review your answer:
1) 1,2,3,4,6,9,12,18,36
2) 1,3,5,9,15,45
3) 1,2,3,5,6,10,15,25,30,50,75,150
4) 1,17

Multiples

Multiples of a number

The product of a number with the natural numbers 1,2,3,4,5,6,7,8,9,10.....

The multiples of 5= 5*1=5, 5*2=10, 5*3=15....etc....

If they ask you "What are the multiples of 24?"
You start with the question
"What is the product of 24 with all the natural numbers?"
That means 24*1=24, 24*2= 48, 24*3= 72...
The multiples of 24 are: 24,48,72, 96.......

Let's Practice:
1) Find the first 5 multiples of 3:
2) Find the first 10 multiples of 11:
3) Find the first 5 multiples of 6:
4) Find the first 10 multiples of 8:

Let's review your answer:
1) 3,6,9,12,15...
2) 11,22,33,44,55,66,77,88,99,100,110...
3) 6,12,18,24,30....
4) 8,16,24,32,40,48,56,64,72,80...

Greatest Common Factor and Least Common Multiple

The **Greatest Common Factor or GCF and the Least Common Multiple** will help you to do factoring and solve equations for the algebra section.

These are two examples of **finding the Greatest Common Factor and the Least Common Multiple for 12 and 24.**

What is the GCF of (12,24) ?

1) Factors of 24 = 1,2,3,4,6,8,12,24
2) Factors of 12= 1,2,3,4,6,12
3) Common Factors: 1,2,3,4,6,12
4) Greatest Common Factor: 12

What is the LCM of (12, 24)?

1) Multiples of 12: 12,24,36,48.....
2) Multiples of 24: 24, 48, 72....
3) Common Multiple(s): 24, 48
4) Least Common Multiple: 24

Let's do some examples to master this concept:

A) Find the GCF of 12 and 36:

The first step is to find the individual factors of 12 and 36.
12: 1,2,3,4,6,12
36: 1,2,3,4,6,9,12,18,36
You can see the COMMON FACTORS ARE (1,2,3,4,6,12)
12: **1,2,3,4,6,12**
36: **1,2,3,4,6**,9,**12**,18,36
The **GREATEST COMMON FACTOR IS 12**

B) Find the GCF of 100 and 25:

The first step is to find the individual factors of 100 and 25.
100: 1,2,4,5,10,20,25,50,100
25: 1,5,25
You can see the COMMON FACTORS ARE (1,5,25)
100: **1**,2,4,**5**,10,20,**25**,50,100
25: **1,5,25**
The **GREATEST COMMON FACTOR IS 25**

C) Find the LCM of 12 and 36
The first step is to find multiples of 12 and 36
12: 12,24,36,48,60,72,84,96,108,120…
36: 36,72,108,144……
You can see the COMMON MULTIPLES ARE SO FAR (36,72,108)
12: 12,24,**36**,48,60,**72**,84,96,**108**,120…
36: **36,72,108**,144……
You need the LEAST one from the list, that is **36**
The **LEAST COMMON MULTIPLE 36**

23

> **These are shortcuts to find the GCF and LCM faster** ☺

Greatest Common Factor (GCF)
Find the GCF of 60 and 45

STEP ONE: First do the prime factorization of 60 and 45
$$60 = 2^2 * 3 * 5$$
$$45 = 3^2 * 5$$

STEP TWO: Multiply only the **common** factors with the **lowest exponent**
$$= 3 * 5$$
$$GCF = 15$$

Least Common Multiple
Find the LCM of 200 and 60

STEP ONE: Find the prime factorization
$$200 = 2^3 * 5^2$$
$$60 = 2^2 * 3 * 5$$

STEP TWO: Multiply all the **common** and **non-common** factors with the **highest exponent**.
$$2^3 * 5^2 * 3 =$$
$$8 * 25 * 3 = 600$$

Find the LCM of 20 and 30
$$20 = 2^2 * 5$$
$$30 = 2 * 5 * 3$$

LCM of 20 and 30 = $\quad 2^2 * 5 * 3 = 60$

Let's practice:

Find the GCF and LCM of the following numbers:
1) 4 and 20
2) 12 and 30
3) 14 and 45
4) 3,6 and 24
5) 12,14 and 26

Let's check your answers:
1) GCF: 4 LCM:20
2) GCF: 6 LCM:60
3) GCF: 3 LCM:180
4) GCF: 3 LCM:24
5) GCF: 2 LCM:1092

Fractions

When you have a fraction you need to remember, you have a numerator and a denominator.

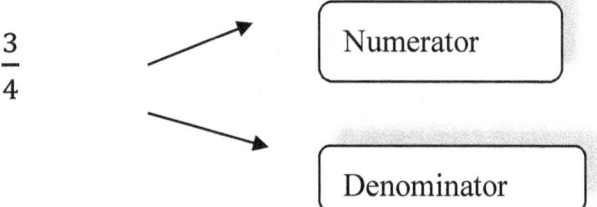

If the numerator is **smaller** than the denominator the fraction is **proper**. If the numerator is **greater** than the denominator the fraction is **improper**.

$$\frac{3}{4} \; proper \quad vs \quad \frac{4}{3} \; improper$$

Mixed Numbers are a whole number and a fraction together. An improper fraction can be converted into a mixed number the following way.

$$\frac{36}{5}$$

Divide the denominator into the numerator (like: 36 ÷ 5). Since 5 goes into 36, 7 times (5 x 7 = 35). The remainder is the difference 36 – 35 = 1. You can now re-write the expression the following way:

$$7\frac{1}{5}$$

Let's practice:

Convert the following fractions into mixed numbers:

a) $\dfrac{36}{5} =$

b) $\dfrac{23}{4} =$

c) $\dfrac{16}{3} =$

D) $\dfrac{8}{3} =$

A) $7\dfrac{1}{5}$ b) $5\dfrac{3}{4}$ c) $5\dfrac{1}{3}$ d) $2\dfrac{2}{3}$

Adding and Subtracting Fractions

When you add or subtract fractions, they need the **same** denominator.

> WHEN THE NUMERATOR AND DENOMINATOR ARE THE SAME NUMBER, IT IS THE SAME AS THE WHOLE NUMBER "1".
> FOR EXAMPLE: 3/3= 1

Let's do some examples

a) $\frac{1}{3} + \frac{2}{3}$ These two fractions have the same denominator.

Add the top and keep the same denominator: $\frac{1+2}{3} = \frac{3}{3} = 1$

b) $\frac{2}{3} + \frac{1}{4}$

In this example, the two fractions do not have the same denominator, $\frac{2(4)}{3(4)} + \frac{1(3)}{4(3)}$ but we can make then equal by multiplying both top and bottom with the opposite denominator (by 3 and 4).

Now you can add the top and keep the denominator: $\frac{8}{12} + \frac{3}{12} = \frac{11}{12}$

.

c) $\frac{3}{7} + \frac{1}{14}$ In this case, you can multiply $\frac{3}{7}$ by 2 to the top and bottom to get 14 as the common denominator.

$\frac{3(2)}{7(2)} + \frac{1}{14} = \frac{6}{14} + \frac{1}{14} = \frac{7}{14} = \frac{1}{2}$.

> **DON'T FORGET TO SIMPLIFY FRACTIONS TO THE LOWEST TERM**

Let's do another example:

$$\frac{1}{3} - \frac{2}{7} = \frac{1(7)}{3(7)} - \frac{2(3)}{7(3)} = \frac{7}{21} - \frac{6}{21} = \frac{1}{21}$$

> **REMEMBER: TO MAKE A WHOLE NUMBER A FRACTION, YOU PUT IT OVER "1"**

Another example:

$\frac{5}{6} + 8$ In this case you have a fraction with a whole number. You can put a 1 on the bottom of 8 to make it look like a fraction.

$\frac{5}{6} + \frac{8}{1}$.

$\frac{5}{6} + \frac{8(6)}{1(6)}$ ⟹ then you can multiply by 6 the top and the bottom

$\frac{5}{6} + \frac{48}{6} = \frac{\mathbf{53}}{\mathbf{6}}$ ⟹ solve the problem

$\frac{53}{6} = \mathbf{8\frac{5}{6}}$ ⟹   or make it a mixed number.

Last example:

$\frac{3}{7} - \frac{1}{3} + \frac{2}{5}$ In this example, we have a combination of three denominators. Just multiply what each denominator needs to be equal to each other. (as long as you do top and bottom)

$\frac{3\,(3)\,(5)}{7\,(3)\,(5)} - \frac{1\,(7)\,(5)}{3\,(7)\,(5)} + \frac{2\,(3)\,(7)}{5\,(3)\,(7)} \implies$ the $\frac{3}{7}$ needs a 3 and a 5.

Then $\frac{1}{3}$ needs a 7 and a 5. Finally, $\frac{2}{5}$ needs a 7 and a 3.

Don't forget, whatever you do to bottom needs to be done to the top.

$\frac{45}{105} - \frac{35}{105} + \frac{42}{105} = \frac{52}{105}$

Finally, you have the same denominator, you can add the top.

Let's Practice!

1) $\frac{1}{3} + \frac{2}{7}$ 2) $1 - \frac{3}{5}$ 3) $\frac{7}{5} + \frac{2}{7}$

4) $4 - \frac{2}{3}$ 5) $6 - \frac{7}{5} + \frac{2}{3}$ 6) $\frac{1}{5} + \frac{1}{2} - \frac{7}{10}$

Answers: 1) $\frac{13}{21}$, 2) $\frac{2}{5}$, 3) $\frac{59}{35}$, 4) $\frac{10}{3}$, 5) $\frac{79}{15}$, 6) 0

Multiplying and Dividing Fractions

When you multiply or divide fractions you don't need to have the same denominator.
Let's do some examples.

$\frac{1}{3} \times \frac{2}{5}$ Just multiply top to top and bottom to bottom.

$= \frac{2}{15}$

Another Example:

$\frac{2}{7} \times \frac{3}{5} \times \frac{1}{8} = \frac{(2)(3)(1)}{(7)(5)(8)} = \frac{6}{280} = \frac{3}{140}$

Let's do another example:

$3 \times \frac{2}{5}$ If you have a whole number with a fraction, just make the whole number a fraction by adding a "1" to the bottom.

$\frac{3}{1} \times \frac{2}{5} = \frac{6}{5}$ or $1\frac{1}{5}$

When you divide fractions, the procedure is almost the same but you need to do one more step.

$\frac{2}{5} \div \frac{1}{8}$

STEP ONE: Keep the first fraction the same and flip the second fraction

STEP TWO: Switch the divide sign to a multiply sign (/ into X)

$$\frac{2}{5} \times \frac{8}{1}$$

STEP THREE: Solve

$$\frac{2(8)}{5(1)} = \frac{16}{5} \text{ or } 3\frac{1}{5}$$

Let's Practice!

1) $\frac{3}{7} \div \frac{1}{5}$ 2) $6 \div \frac{3}{7}$ 3) $\frac{2}{5} \div 7$

4) $\frac{6}{5} \div \frac{1}{8}$ 5) $\frac{1}{7} \div \frac{1}{3}$ 6) $\frac{7}{5} \div \frac{1}{8}$

Answers: 1) $\frac{15}{7}$ 2) 14 3) $\frac{2}{35}$ 4) $\frac{48}{5}$ 5) $\frac{3}{7}$ 6) $\frac{56}{5}$

Percentages

If you have 100 units, then 50% is half of 100, that is 50 units
If you have 200 units, then 50% is half of 200, that is 100 units
If you have 30,000 units, then 50% is half of 30,000, that is 15,000 units.

As you can see, percentages are a rate of a whole for every hundred.
Percentages are represented by the symbol %, let's learn the basics of percentages:

> Percentages can be expressed as a **percentage, a fraction or a decimal.**

These are the most common percentages represented as fractions

$$\frac{1}{10} = 10\% \quad \frac{1}{5} = 20\% \quad \frac{3}{10} = 30\%$$

$$\frac{2}{5} = 40\% \quad \frac{1}{2} = 50\% \quad \frac{3}{5} = 60\%$$

$$\frac{7}{10} = 70\% \quad \frac{8}{10} = 80\% \quad \frac{9}{10} = 90\%$$

Convert a percent into a decimal

Convert 30% into a decimal
1) Write the number without a %
30
2) Divide the number by 100
30/100=0.30
3) Or you can also move to the LEFT TWO TIMES

Convert a decimal into a percent

Convert 0.25 into a percent
1) Multiply the decimal by 100
0.25*100= 25
2) Place the sign % next to the number
25%
3) You can also move to the RIGHT TWICE to make it into a percent.

Convert a fraction into a percent

Convert 3/5 into a percent
1) Do the division or convert 3/5 into a decimal
3/5=0.6
2) Multiply it by 100
0.6*100=60
3) Add the % symbol to your answer
60%

> **Convert a percent into a fraction**
>
> Convert 40% into a fraction
> 1) Write the number without the symbol %
> 40
> 2) Divide it by 100
> 40/100
> 3) Simplify the fraction to its lowest term.
> 40/100=2/5

Let's do some examples to understand how percentages are represented

1. 20% is also 0.2 and 20/100=1/5
2. 34.5% is also 0.345 and 34.5/100=69/200
3. 35% is also 0.35 and 35/100=7/20
4. 234% is also 2.34 and 234/100= 117/50

Let's practice:

Convert the following percentages into decimals and fractions:
1) 23%
2) 36%
3) 12.45%
4) 4.78%
5) 0.4%
6) 123%

Let's check your answers:
1) 0.23, 23/100
2) 0.36, 9/25
3) 0.1245, 249/200
4) 0.0478, 239/5000
5) 0.004, 1/250
6) 1.23, 123/10000

Convert the following decimals into percentages:
1) 0.034
2) 0.34
3) 0.52
4) 3.45
5) 0.00567
6) 34.67

Let's check your answers;
1) 3.4%
2) 34%
3) 52%
4) 345%
5) 0.567%

6) 3467%

Convert the following fractions into decimals and then percentages:

1) 3/8
2) 1/9
3) 3/4
4) 13/20
5) 11/20
6) 21/50

Let's check your answers;
1) 0.375, 37.5%
2) 0.1111, 11.11%
3) 0.75, 75%
4) 0.65, 65%
5) 0.55, 55%
6) 0.42, 42%

Percentages Increase and Decrease:

Let's say you want to buy a phone that is $105 dollars on Monday, and then on Friday the same phone is selling for $84 dollars, you get really happy and wonder what was the percent decrease from 105 to 84.

It is really easy to do by using this formula:

$$Percent\ Change\ \% = \frac{Change}{Original} \cdot 100$$

Percent Change= (105-84)/105= 0.21 or 21%
The phone decreased 21% of its original price.

Let's do another example:

Peter's salary went from $100 to $160. What is the percent increase?

$$Percent\ Change = \frac{160-100}{100} = \frac{60}{100} = 0.6$$

That is 0.6 times 100 is 60%

Peter's salary went from $125 to $90. What is the percent decrease?

$$Percent\ Change = \frac{125-90}{125} = \frac{35}{125} = 0.28$$

That is 0.28 times 100 is 28%

> 1) The original always goes on the bottom.
>
> 2) The percent decrease from 100 to 70 is not the same as the percent increase from 70 to 100!
>
> Let me show you:
> Percent Decrease from 100 to 70: (100-70)/100= 0.3 or 30%
> Percent Increase from 70 to 100: (100-70)/70=0.4285 or 42.85%

Let's Practice:

Find the percent change (Increase or decrease) of the following numbers.

1) What is the percent decrease of 500 to 400?
2) What is the percent decrease of 600 to 378?
3) What is the percent increase of 400 to 500?
4) What is the percent increase of 100 to 170?
5) What is the percent decrease of 170 to 100?

Let's check your answers:

1) Percent Decrease: (500-400)/500= 0.20 or 20%
2) Percent Decrease: (600-378)/600= 0.37 or 37%
3) Percent Increase: (500-400)/400=0.25 or 25%
4) Percent Increase: (170-100)/100=0.70 or 70%
5) Percent Decrease: (170-100)/170=0.4118 or 41.18%

Percentages word problems

Percentages word problems are very common; let's do examples of several common percentages word problems.

In order to find percentages you need this formula

$$\frac{is}{of} = \frac{Percent}{100}$$

Let's do an example!!

1) What is 30% of 120?

You are looking for the part (is) of the problem, you have the percentage and the whole (of), plug the values in to the formula and "X" is what you are looking for.

X/120=30/100
Cross-multiplying both sides:
X(100)=120(30)
X(100)=3600
X= 3600/100= 36
36 is 30% of 120

2) 12 is what percent of 20?

Apply the formula, you need the percent, make percent "X" and solve for it.

$12/20 = X/100$

Cross-multiplying both sides:

$12(100) = 20X$

$1200 = 20X$

$X = 1200/20 = 60$

12 is 60% of 20

3) 25% of what number is 300

Using the formula, you need to find the whole(of)

$300/X = 25/100$

$300(100) = 25(X)$

$30000/25 = X$

1200

25% of 1200 is 300

Let's Practice:
1) 28% of what number is 560.
2) 49 is what percent of 350.
3) What is 12% of 1250.
4) 30% of what number is 966.
5) What is 12.75% of 250.

Answers: 1) 2000 2) 14% 3) 150 4) 3220 5) 31.875

Let's check your answers:
1) $560/X = 28/100$ Cross-multiplying
$560(100) = 28X$
$56000/28 = x$

X=2000

2) $49/350 = X/100$
$49(100) = 350X$
$4900/350 = x$
X=14%

3) $X/1250 = 12/100$
$X(100) = 1250(12)$
$X = 15000/100$
X=150

4) $966/X = 30/100$
$966(100) = X(30)$
$X = 96600/30 = 3220$

5) $X/250 = 12.75/100$
$100(X) = 250(12.75)$
$100(X) = 3187.5$
$X = 3187.5/100$
X=31.875

Ratios

If you need to compare two quantities with the same unit, you can use ratios.

Ratios

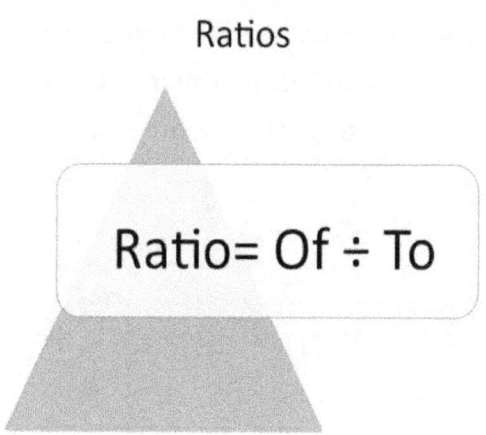

Let's say you have a basket with 2 apples and 5 oranges, then you can say:
The number of apples to the number of oranges can be written the following ways:

1) *As a fraction:* $\frac{2\ apples}{5\ oranges}$

2) 2 to 5

3) 2:5

BE CAREFUL, IF YOU WANT TO REPRESENT THE RATIO, then you need to find the whole, in this case how many fruits in total you have.
Whole= 2 apples + 5 Oranges= 7 fruits in total

The ratio of apples to the whole (both fruits)
2/(2+ 5) =2/7
The ratio of oranges to the whole (both fruits) 5/(2+5)= 5/7

Let's do an example:

The ratio of cats and dogs in a non-killer animal shelter is 4:5.
What does this mean?
You need to make a fraction from each animal this way.

$$\frac{4}{5+4} = \frac{4}{9}$$ from 9 animals, 4 are cats

$$\frac{5}{5+4} = \frac{5}{9}$$ from 9 animals, 5 are dogs

Let's do another example:

In a party of 24 teenagers, the ratio of girls and boy is 5:7.
How many girls are in the party?
In this problem, we need to figure out the number of girls and boys.

Let's do the fraction of the number of girls at the party:
First let's figure out the ratio of girls and boys at the party:
Girls ratio: 5/(7+5) = 5/12
Boys: 7/(7+5) = 7/12
To find the how many girls, then multiply the total number of teenagers to the ratio of girls:
Girls: 24(5/12) = 10

Boys: 24(7/12) = 14

There are 10 girls and 14 boys at the party

Let's Practice:

1) In a cool party, there are 45 guests, the ratio of boys to girls is 3 to 2, how many girls are in the party? How many boys are in the party?

2) Alison and Rosy are playing a video game, the ratio of the scores is 5 to 9, respectively. If the total score is 84 points, how many points has Rosy scored?

3) A newspaper states that the ratio of students that like rock music is 4:17. If 189 students were surveyed, how many do not listen to rock?

4) A granola bar contains 3:10 grams of sugar to protein. If the granola bar weights 91 grams, how many grams of sugar and protein does it contains?

Let's check your answers:

1) Boys ratio: 3/5
Girls ratio: 2/5
Boys= 45(3/5)= 27
Girls= 45(2/5)=18
There are 27 boys and 18 girls in the party.

2) Alisson ratio: 5/(5+9)= (5/14)* 84= 30 points
   Rosy ratio: 9/(5+9)= (9/14)*84= 54 points

3) If 4/7 like to listen to rock, then the complement or 1-4/7 = 3/7 do not listen to rock.

(3/7)*189=81, that means 81 students do not like to listen to rock

4) Sugar ratio: 3/(3+10)= (3/13) 91= 21 grams of sugar

Protein ratio: 10/(3+10)= (10/13)91= 70 grams of protein

Rates and proportions

Rates can relate two different units.
For example: miles per hour, dollars per pound, litters per minute
Proportions compares two ratios

Common Rates examples:
1) Vera drives 60 miles every hour
2) A chocolate store sells 15 dollars of chocolate per pound
3) The hose fills the pool 0.23 gallons per minute

Proportions

Proportions are used to compare two ratios.
For example:
If there are 60 minutes in 1 hour,
how many minutes are in 4 hours?

$$\frac{60 \; minutes}{1 \; hour} = \frac{X}{4 \; hours}$$

Solve for X by:

$$60(4) = X$$
$$240 = X$$
$$X = 240 \; minutes$$

Let's solve the problem:
Kristine can eat 4 cupcakes in 10 minutes, how many cupcakes can she eat in 180 minutes?

$$\frac{4 \; cupcakes}{10 \; minutes} = \frac{X}{180 \; minutes}$$

$$4(180) = 10X$$
$$720 = 10X$$
$$X = \frac{960}{10} = 72 \text{ cupcakes}$$

That is a lot of cupcakes :)

1) A machine can produce 12 boxes every 4 hours. How many boxes can the same machine produce in one day?

2) Dorothy reads 30 pages every 5 minutes. If a book has 165 pages, in how many minutes can she finish the book?

3) What is the ratio of 6 minutes to 6 hours?

4) Paul types 36 words per minute. How many words can he type in 4¾ minutes?

5) A machine can produce 360 toys in 12 hours. At this rate, how many toys can the machine produce in 6 minutes?

Let's check your answers:
1) There are 24 hours in a day, then:
12/4= x/24
12(24) =4x
x= 72 boxes

2) 30/5= 165/x
30(x) =165(5)
x= 27.5 minutes

3) 6 minutes to 6 hours need to be in the same unit. Let's make the hours into minutes

$$\frac{6\ minutes}{6\ hours} * \frac{1\ hour}{60\ minutes} = \frac{1}{6}$$

4) 36/1= x/4.75
(36*4.75) = x
x= 171 words

5) $\frac{360}{12(60)}$ =(x/5) (remember that you need to make 12 hours into minutes, that is why we multiply by 60)

360/720= x/6
360(6)=720x
x=3 toys

Sets and the Venn Diagram

A set is a collection of numbers or other objects.

Inside the set {} you have {elements}.

For example, set U has 4 elements.

Set U
U= {1,2,5,7}

Let's Practice

If you have two sets A and B:

A= {1,4,5,6,7}

B= {1,3,4,5,6,7,8,11,12,14}

A∪B means the **Union** for A and B, this means
ALL COMMON AND NON-COMMON ELEMENTS

A∩B means the **Intersect** for A and B, this means
ONLY the COMMON ELEMENTS

- The **union** of the two sets **A U B**, is the set of all the elements that are in A, B or both.

  The union of the sets is: A U B = {1,3,4,5,6,7,8,11,12,14}

- The **intersect** of the two sets A∩B, is the set of all elements that are only common in A and B.

  The intersection of the two sets $A \cap B$ = {1,4,5,6,7}

- The **complement** of a set, is everything that does not belong to that set. The notation is A', or B'. Sometimes the notation looks like this ~A, or ~B.

  The complement of set A, A'= {3,8,11,12,14}

Venn Diagram

The Venn diagram is a graphic representation of the sets.

For example:

Set U has 13 elements, then you have two sets, A and B.
U= {1,2,3,4,5,6,7,8,9,10,11,12,13}

A= {1,2,3,4,5,7,9}

B= {3,4,5,6,8,10}

Create a Venn Diagram representing the UNION and INTERSECTION

The **union** of the two sets is A U B= {1,2,3,4,5,6,7,8,9,10}

The **intersection** of the two sets is $A \cap B$ = {3,4,5}

The Venn diagram shows the graphic relation for sets A and B

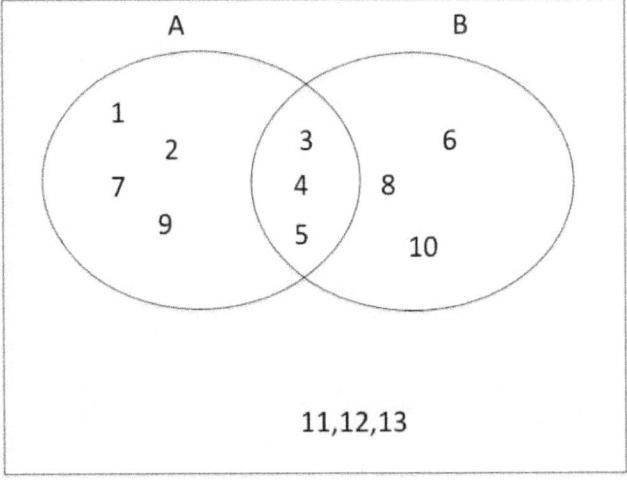

Let's Practice

The following Venn Diagram represents 3 sets A, B and C.

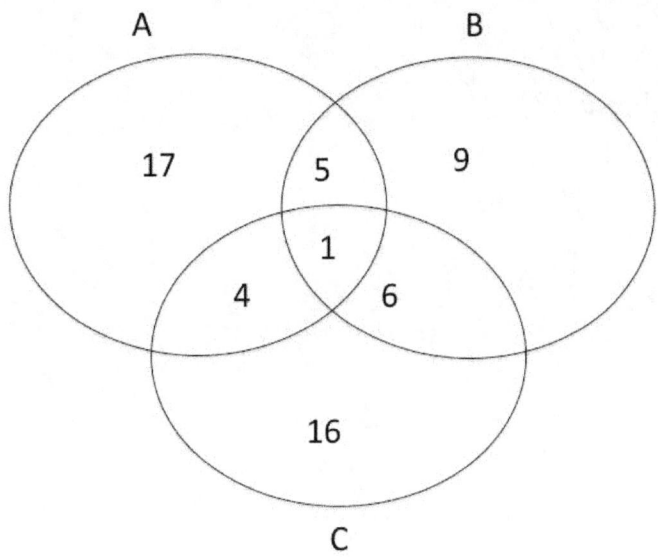

Find the following:

1) Set A

2) Set B

3) Set C

4) A∪B

5) A'

6) A ∩ B

7) A ∩ B ∩ C

8) A ∩ C

9) A' U C

10) (A ∩ B) ∩ 'C

Let's check your answers

1) {1,4,5,17}
2) {1,5,6,9}
3) {1,4,6,16}
4) {1,4,5,6,9,17}
5) {6,9,16}
6) {1,5}
7) {1}
8) {1,4}
9) {6,9,16}
10) {5}

PROBABILITY

The probability of an event is determined by the following formula:

P(E)= #Success/Possible outcomes

For example, the probability when rolling a fair die and getting a 6 is:

P(6)= 1/6, that is there is 1 success (YOU CAN ONLY GET ONE 6) with 6 outcomes (1,2,3,4,5,6)

The **probability of an event not happening** is:

P(not happening) = 1 - P(E)

The probability of rolling a fair die and not getting a 6 is:

P(not 6) = 1- (1/6)= 5/6

In the problems below find the probability when rolling a fair dice and getting...

1) The number 2

2) The number 3

3) The number 7

4) A number less than 3

5) A number greater than 3

6) A number greater than 5

7) An odd number

8) An even number

Answers:
1) 1/6
2) 1/6
3) 0
4) 1/3
5) 1/2
6) 1/6
7) 1/2
8) 1/2

Probability of independent/dependent events

When two events are **independent**, the fact that A occurs **does not affect** the probability of B occurring.

You need to use this formula:

> $P(A \text{ and } B) = P(A) \cdot P(B)$
>
> **Replacement means the probability is independent**

For example:

In a bag of marbles, there are 3 red and 5 blues. If two marbles are chosen from the jar with **replacement.**

What is the probability of the two marbles being red?

$P(red) = 3/(3+5) = 3/8$

Since you are choosing the marble and then putting it back in the bag, the probability is **INDEPENDENT**
$P(red \text{ and } red) = (3/8)(3/8) = $ **9/64**

When two events are **dependent**, the probability of A affects the probability of B

When two events are dependent use this formula:

> **P(A and B)= P(A) * P(B after A happened)**
>
> This happens when there is **NO REPLACEMENT**

If I have the same problem but **WITHOUT REPLACEMENT**
In a bag of marbles, there are 3 red and 5 blues. If two marbles are chosen from the jar **without replacement**.

What is the probability of the two marbles being red?

In this case the probability is **DEPENDENT**, that is once the red marble is chosen the first time P(red): 3/8, the second time there are only 2 out 3 red marbles, and the bag only has 7 marbles, therefore the probability of choosing two red marbles is:

P(red and red) = (3/8)(2/7) = **3/28**

Let's Practice

1) In a bag of marbles, there are 4 white marbles, 5 red marbles and 7 black marbles. Two marbles are drawn, with replacement. What is the probability of getting two white marbles?

2) In a bag of marbles, there are 4 white marbles, 5 red marbles and 7 black marbles. Two marbles are drawn, with replacement. What is the probability of getting two red marbles?

3) In a bag of marbles, there are 4 white marbles, 5 red marbles and 7 black marbles. Two marbles are drawn, without replacement. What is the probability of getting two red marbles?

4) Two cards are chosen at random from a deck of 52 cards without replacement. There are 4 kings in a deck of cards. What is the probability of getting two kings?

5) Two cards are chosen at random from a deck of 52 cards with replacement. There are 4 kings in a deck of cards. What is the probability of getting two kings?

6) On a math test, 7 out of 20 students got an B. If three students are chosen at random without replacement, what is the probability that all three got an B on the test?

1) (4/16)(4/16) = (16/256)=1/16
2) (5/16)(5/16)= 25/256
3) (5/16)(4/15)=1/12
4) (4/52)(3/51)= 1/221
5) (4/52)(4/52)= 1/169
6) (7/20)(6/19)(5/18)=7/228

Probability of Mutually Exclusive Events (Disjoint)

Two events are mutually exclusive when **they can NOT happen** at the **same time**, for example: getting heads or tails, being female or male, being democrat or republican.

> **Mutually Exclusive:**
> Two events *cannot* happen at the same time
> **P(A or B) = P(A) + P(B)**

Let's practice

1) The probability of rolling a fair die and getting a 2 or 3

2) The probability of rolling a fair die and getting a 1 or 4

3) The probability of tossing a coin and getting a tail?

4) If the probability of Maria winning a game is 1/4 and the probability of Juan winning the same game is 3/7. Find the probability that either Maria or Juan can win the game?

5) If the probability of Maria winning a game is 1/4 and the probability of Juan winning the same game is 3/7. Find the probability that NEITHER Maria or Juan can win the game?

6) If P(A) = 4/5 and P(B) = 1/16, and the two events are mutually exclusive. What is the probability that P(A or B)?

7) From a bag containing 6 white marbles, 2 black marbles, and 12

red marbles, if 1 marble is drawn. What is the probability that it is either black or red?

8) From a bag containing 6 white marbles, 2 black marbles, and 12 red marbles, if 1 marble is drawn. What is the probability that it is either white or black?

9) From a bag containing 6 white marbles, 2 black marbles, and 12 red marbles, if 1 marble is drawn. What is the probability that it is Neither black or red?

10) From a bag containing 6 white marbles, 2 black marbles, and 12 red marbles, if 1 marble is drawn. What is the probability that it is either red or white?

11) From a bag containing 6 white marbles, 2 black marbles, and 12 red marbles, if 1 marble is drawn. What is the probability that it is Neither red or white?

Let's check your answers

1) 1/3 2) 1/3 3) 1/2 4) 19/28 5) 9/28 6) 69/80 7) 7/10 8) 2/5 9) 3/10 10) 9/10 11) 1/10

Probability of Not Mutually Exclusive Events (Joint)

When two events **are not mutually exclusive, they can happen at the same time**, we use this formula:

> **Not-Mutually Exclusive**
> Two events **Can** happen at the same time
> (The probability of being a woman or a mom)
> P(A or B) = P(A) + P(B) −P(A and B)

For example:

The probability of being a woman and being democrat, the probability of being a sport player and being a male.

Let's do an example:

If P(A)=0.60 and P(B)=0.30 are independent and not mutually exclusive, then find P(A or B)

1) Find P(A and B) first
P(A and B) = (0.6)(0.30) = 0.18

2) Now apply the formula

P(A or B) = P(A) + P(B) - P(A and B)

P(A or B) = 0.60 + 0.30 - 0.18 = 0.72

Let's practice

The table shows the distribution of a group of 40 students by gender and sports

	Basketball	Soccer	Tennis
Male	3	8	4
Female	9	10	6

If one student is randomly chosen, then:

1) What is the probability of being a female?

2) What is the probability of being a male?

3) What is the probability of being a male or a tennis player?

4) What is the probability of being a male or a soccer player?

5) What is the probability of being a female or a basketball player?

6) What is the probability of not being a female or a basketball player?

7) Given that a female is chosen, what is the probability of being a tennis player?

8) Given that a female is chosen, what is the probability of not being a soccer player?

Let's check your answers

1) What is the probability of being a female? = (9+10+6)/40 = 25/40 = **5/8**

2) What is the probability of being a male? = (3+8+4)/40 = 15/40 = **3/8**

3) What is the probability of being a male and a tennis player?
P(male or tennis) = P(male) + P(tennis) - P(male and tennis)
P(male or tennis) = (3/8) + (10/40) - (4/40) = **21/40**

4) What is the probability of being a male or a soccer player?
P(male or soccer) = P(male) + P(soccer) - P(male and soccer)
P(male or soccer) = 3/8 + 18/40 - 8/40 = **5/8**

5) What is the probability of being a female or a basketball player?

P(female or a basketball player) = P(female) + P(basketball) - P(female and basketball)
P(female or a basketball player) = 5/8 + 12/40 - 9/40 = **7/10**

6) What is the probability of not being a female or a basketball player?

P(not being a female or basketball player) =
1 - P(female or basketball player)
1 - 7/10 = **3/10**

7) Find the total of females first, (9+10+6)= 25, then a tennis player is 6, finally **6/25**

8) Find the total of females first, (9+10+6)= 25, then find the tennis and basketball players (9+6) =15, then 15/25 **= 3/5**

Factorial Notation

The factorial of N! is the product of all positive integers from N to 1.
For example:

> $0! = 1$
> $1! = 1$
> $2! = 1*2 = 4$
> $3! = 1*2*3 = 6$
> $4! = 1*2*3*4 = 24$

We can do operations with factorials, for example:

- $5! - 4! = 120 - 24 = 96$

- $6!/5! = (1 \cdot 2 \cdot 3 \cdot 4 \cdot 5 \cdot 6)/(1 \cdot 2 \cdot 3 \cdot 4 \cdot 5) = 6$

Let's practice!

1) $3!$

2) $4! - 3!$

3) $(4-2)!$

4) $6!/3!$

5) $1!$

Answers: 1) 6 2) 18 3) 2 4) 120 5) 1

Counting Methods

If you have "m" ways of doing one thing and then "n" ways of doing another then m*n is how many ways you can do both, this sentence represents the fundamental principle of counting.

For example:
Alex can choose between 4 appetizers, 5 main meals, and 3 desserts. How many different meals can he order?

Counting: 4*5*3= 60 meals

Combination and Permutation

When you have to select items from a collection, you need to do a combination or a permutation, depending on if the order matters.

Combination and Permutation

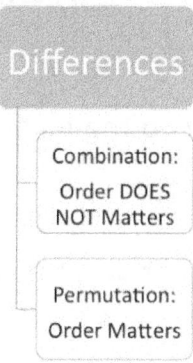

Formulas:

n is the total (biggest number) and r is the sets needed (the smallest number)

Combination:

$$C^n_r = \frac{n!}{(n-r)!\,r!}$$

Permutation:

$$P^n_r = \frac{n!}{(n-r)!}$$

For example: you have 4 ice cream flavors:
Vanilla (V), Chocolate (C), Orange (O), Mint (M).
How many ways can you chose 2 flavors out of the 4?

n=4

r=2

$$C^4_2 = \frac{4!}{(4-2)!\,2!} = \frac{4*3*2*1}{2*2} = 6 \text{ ways}$$

You can do the following ways:

VC, VO, VM, CO, CM, OM, that is 6 ways.

Does it matter if you eat vanilla first and then chocolate?
NO, because VC is the same as CV, the order DOES NOT MATTER, then **you have to use a combination.**

Let's do a permutation example:

You have 6 candidates running for the board of an association. The board requires a President (P), Vice-president (VP) and a Treasurer (T). How many ways can you select the board?

There are 6 candidates and 3 positions, however here ORDER MATTERS because there is a ranking.

Let's say you have A, B, C, D, E, F to run for the board.

If you first chose ABC, that is not the same as CBA because A is the president in the first round and C is the president in the second round. ORDER MATTERS

$$P^6_3 = \frac{6!}{(6-3)!} = \frac{6*5*4*3*2*1}{3*2*1} = 120$$

In fact, there are 120 ways to select 3 people out of the 6 to be on the board:

ABC, ACB, BAC, BCA....

THIS IS A PERMUTATION PROBLEM

Let's practice:

Calculate the following combinations and permutations:

1) C_4^7

2) C_6^9

3) C_8^{10}

4) P_5^7

5) P_4^6

6) 6 girls have volunteered to work for the local charity building. How many ways can 3 girls be selected?

7) How many ways can JoJo pick 4 types of ice cream from 7 flavors?

8) How many ways a president, vice-president, and a secretary can be selected from a group of 5 people?

9) How many ways can 6 people be selected 1st, 2nd and 3rd place in a competition?

10) How many 3-digit positive integers are even and do not contain the digit "6"?

11) A jar has 9 pencils inside, 5 are to be removed. How many different sets of 5 pencils could be removed?

12) From a group of 5 boys and 9 girls, a professor must randomly select 3 boys and 2 girls for a study. How many ways can the study group be selected?

1) 35

2) 84

3) 45

4) 2520

5) 360

6) $C_3^6 = 20$

7) $C_4^7 = 35$

8) $P_3^5 = 60$

9) $P_3^6 = 120$

10) The first digit could be 1,2,3,4,5,7,8,9, that is 8 choices, the second digit can be also 1,2,3,4,5,6,7,8,9, that is also 8 choices since you can repeat the digit. Finally, for the number to be even the last digit can be 0,2,4,6,8, that is 5 choices.

8*8*5= 320

11) $C_5^9 = 126$

12) $C_3^5 C_2^9 = 360$

Fundamentals of Statistics

The test will have some questions referring to statistics.

Let's learn the basic concepts.

What is a population?

This is a whole data to be studied. Since sometimes the population is too big, then we use a sample.

What is a sample?

A sample is a subset (or just a piece) of the population.

For example:

Population: All the celebrities in television.

Sample: the celebrities that work for one show.

Population: All the students in a school.

Sample: the students taking Algebra 1.

Type of data

Quantitative data: any data that can be measure or count.
Continuous: This data you can **measure**, for example: the height of people, the weight of a person, the speed of car, and the temperature of an object. **Discrete:** This data you can **count**, for example: the number of students or the number of computers.

Qualitative data: any data that falls into categories. For example: the political affiliation (Democrat, Republican or Independent), the ranking of a movie (Good or Bad), gender (female, male).

Once you have your data, then you can do an **experiment** by collecting and analyzing the data using a ***treatment and a control*** group, or you can do an **observational study** by just collecting opinions or surveys.

Once you select a ***Random sample***, this means any different sample size has **an equal chance** of being selected, then you will avoid being **Bias,** this means you can **cheat the results** by selecting your preferred sample. **For example:** if you want to determine who is the hottest kid in your classroom and you only ask your friends, most likely they will say "You are", that is **being bias**, you should get a hat and collect a survey (in a random manner) from your whole classroom to make that decision.

Once the data has been collected then we can represent it by using either a **dot-plot or a stem-and-leaf display**, these are the **most common graphs** you will see on the test.

Let's do a Dot-Plot:

Let's say you have the following data:

1,3,6,9,6,5,3,6,5,4

1) Organize the numbers from **smallest to greatest**:
1,3,3,4,5,6,6,6,9

2) Now you can see the data in your **dot-plot**:

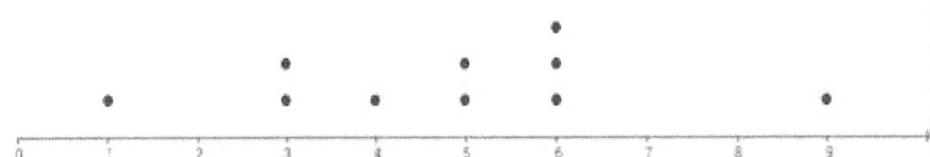

Let's do now a stem-and-leaf graph.

Let's say you have the following data:
12,34,16,29,26,15,43,46,56,47,12,43

1) Organize the numbers from **smallest to greatest**:
12,12,15,16,26,29,34,43,43,46,47,56

The stem will have the first digit and then the leaf the rest.

Stem	Leaf
1	2 2 5 6
2	6 9
3	4
4	3 3 6 7
5	6

Central Measurements

There are 3 central measurements you need to know and understand their differences: **The mean, the median and the mode**.

Mean: it is an arithmetic average; you divide the sum of the measurements by the number of measurements. ***Outliers*** affect the mean, these are points **extreme** (usually too large or small) relative to the data.

For example: Find the mean of the following data:

3,5,6,8,0,20

Mean: (3+5+6+8+0+20)/6=**7**

As you notice the outlier is the number **20, since it is an extreme data from the set.**

Let's find the mean without the outlier (remove the number 20)

Mean: (3+5+6+8+0)/5= **4.4**

Median: it is the middle number when the data is arranged in ascending or descending order. The median **is not affected by outliers** since the middle number it the one that matters.

Let's do an example:

Find the median of the following data:

13,5,16,9,10

1) Arrange the data in ascending order: 5,9,**10**,13,16
2) The median is the number 10 (middle number)

Mode: It is the value of the data most repeated.

The mode is it **not affected by outliers**.

Find the mode of the following data:

4,6,10,100,24,6,7,6

The mode is 6 since it is repeated 3 times.

The 5-Number Summary

The 5-number summary is represented by the minimum value, quartile 1, median, quartile 3 and the maximum value of a data set.

> **The 5-Number Summary:**
>
> Minimum, Quartile 1, Median, Quartile 3, Maximum
>
> The 5 number Summary is represented by the box plot

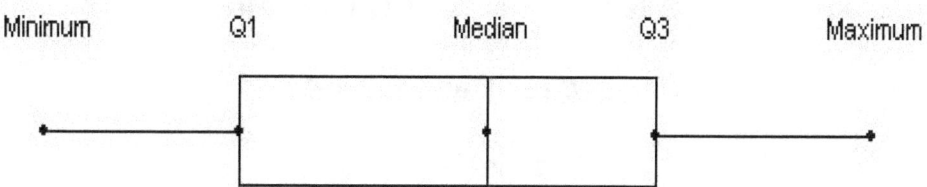

Let's first talk about the quartiles:

Quartile 1: this is the 25th percentile of a data set.

Quartile 2: this is the 50th percentile of a data set (Median)

Quartile 3: this is the 75th percentile of a data set.

Let's do an example:

Find the Q1 and Q3 of the following data: 4,6,10,7,9,1

1) First rearrange data in **ascending order**: 1,4,**6,7**,9,10
2) Find the **Median**, in this case the median is (6+7)/2= 6.5.
3) Pick the values **BELOW** the median 1,4,6 1,**4**,6

Q1 is the middle of these values: **Q1=4**

For Q3, you do the same, but you pick the values **ABOVE THE MEDIAN** 6.5, that is: 7,9,10

7,**9**,10

Q3 is the middle of these values.

Q1= 4 and Q3= 9

Another example:

Find the Q1 and Q3 of the following data:

10,34,56,80,90,56,1,9

1) Arrange them in **ascending order**:

1,9,10,**34,56**,56,80,90

2) Find the **Median** (34+56)/2= 45

3) The values **BELOW** the median are:

1,**9,10**,34,

the middle number of these values (9+10)/2= 9.5

Q1= 9.5

4) The values ABOVE the median are:
5) 56,*56,80*,90, the middle values are (56+80)/2= 68

Q3=68

Now that you understand the quartiles, let's do a five-number summary:

Find the 5-number summary and draw a box plot:

Data: 10,6,8,22 ,14,19,3

1) Rearrange the data in ascending order: 3,6,8,10,14,19,22

2) Find the Minimum, Q1, Median, Q3, and Maximum Minimum: 3 Median: (It is easier to do the median first)= 10
3,6,8,**10**,14,19,22

Q1= 6   Q3= 19

Maximum: 22

THE Box Plot shows the graphic representation of the results:

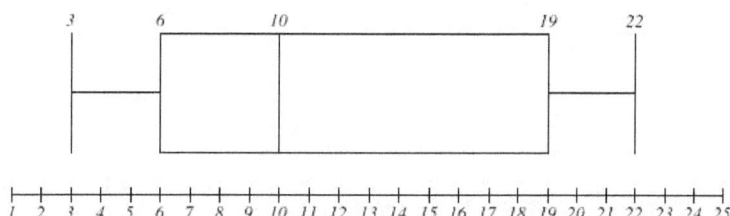

Measurements of Dispersion

There are 4 measurements of dispersion: the interquartile range, the range, the standard deviation and the variance.

These measurements show how **DISPERSE** is the data relative to the mean.

Interquartile Range: IQR= Q3-Q1

Range= Maximum value – Minimum Value

Standard Deviation (sample):

$$s = \sqrt{\frac{\sum(x - \bar{x})^2}{n - 1}}$$

Variance:

$$s^2 = \frac{\sum(X - \bar{X})^2}{n - 1}$$

Let's do an example:

Find the measurements of dispersion for the following data:
1,3,5,6,7

IQR= Q3-Q1= 6.5-2= **4.5**

Range= 7-1= **6**

Standard deviation: **2.15**

1) Find the mean first: Mean= (1+3+5+6+7)/5= 4.4

2) Find the difference square between each value and the mean divide it by n-1 and then square root it.

Standard Deviation= $\sqrt{\dfrac{(1-4.4)^2+(3-4.4)^2+(5-4.4)^2+(6-4.4)^2+(7-4.4)^2}{5-1}}$ = **2.41**

Variance= 2.41^2= **5.8**

> If the data is **symmetrical** (no outliers), then use **the mean and standard deviation**
>
> If the data is **asymmetrical** (with outliers), then use the **median and Interquartile Range**

Let's practice:

For the following data find the **Mean, Median, Mode, Range, IQR and standard deviation. Also draw a box plot.**

a) 2,5,10,4,5,1,2,5

b) 1,0,9,9,1,20

c) 4,6,8,8,8,8,10,11,12

d) -3,5, -10,1, -11

Let's check your answers:

a) Mean=4.25 Median=4.5 Mode=5 Range=9 IQR=3 SD=2.82

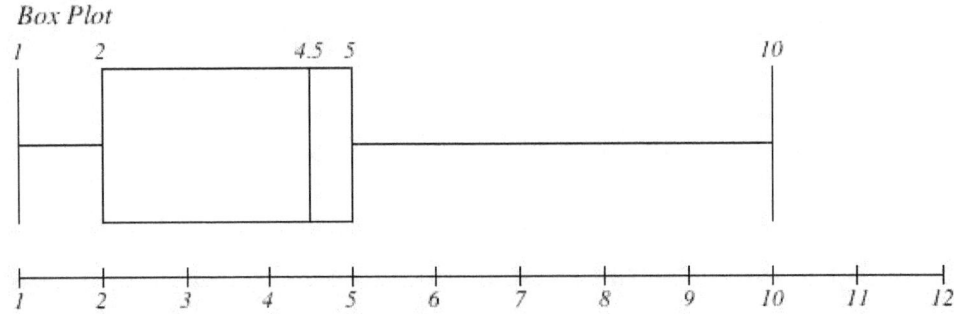

b) Mean=6.67 Median= 5 Mode=9 Range= 20 IQR=8 SD=7.71

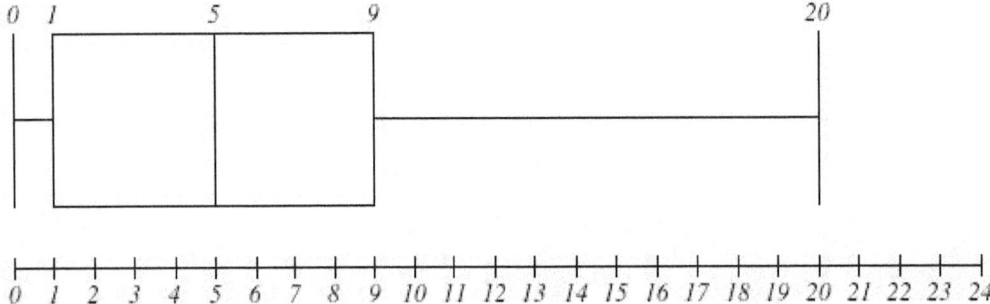

c) Mean=8.33 Median= 8 Mode=8 Range= 4 IQR=3.5 SD=2.45

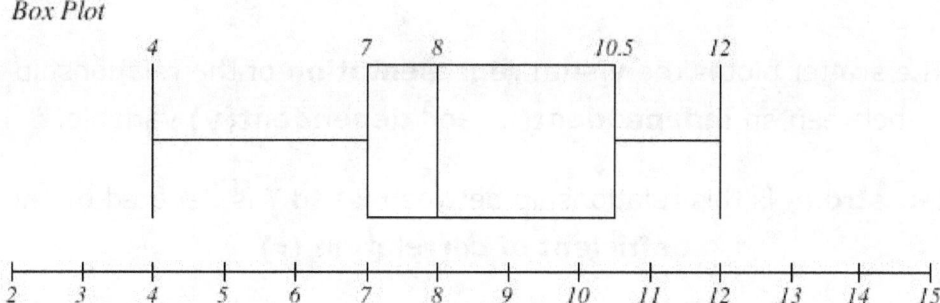

D) Mean=-3.6 Median=-3 Mode=None Range= 16 IQR=13.5 SD=6.91

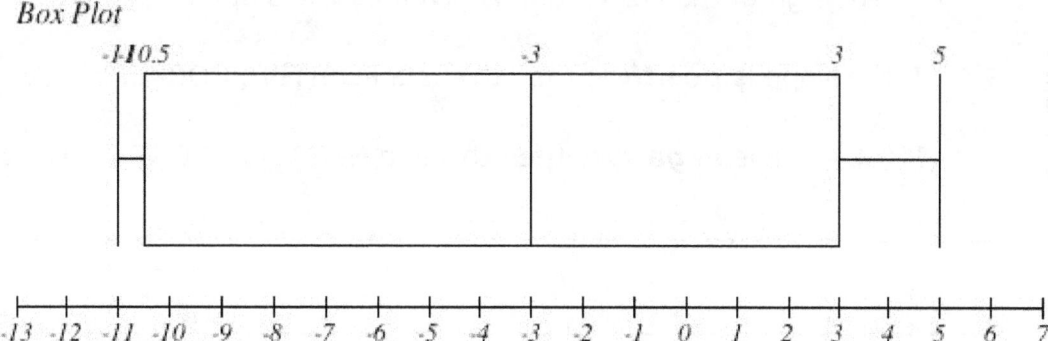

Scatter Plots and Correlation

The scatter plot is the **visual representation** of the relationship between an **independent(x)** and **dependent(y)** variable.

How **strong** is this relationship between x and Y is denoted by the **coefficient of correlation (r)**

Correlation Sign

The sign of the correlation is given by the slope of the line.

If the slope is *positive*, then this is a POSITIVE CORRERLATION

If the slope is *negative*, then this is a NEGATIVE CORRELATION

Correlation strength

The higher the correlation percentage, the stronger the correlation between X and Y

For example: smoking cigarettes has a **strong correlation** of getting cancer, the hours you study for the EOC has a strong correlation with getting a better grade. The value of the coefficient of correlation ranges from -1 to 1. The **closer the dots** are to the line (regression line), the **stronger** the correlation between the two variables

The following graphs show the graphs for correlations:

Positive Correlation

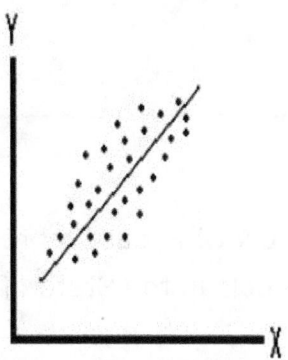

Negative Correlation

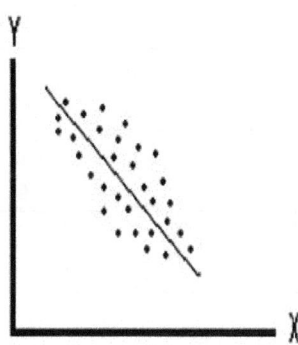

No Correlation

Let's say you have the following data representing the amount of online schools in the State of Florida.

Year	Number of Online Schools
2009	3
2010	6
2011	9
2012	10
2013	11
2014	12
2015	13

Let's plug the values into a scatter plot where **x** is the independent variable (**years**) and **y** is the dependent variable (**number of online schools in Florida**)

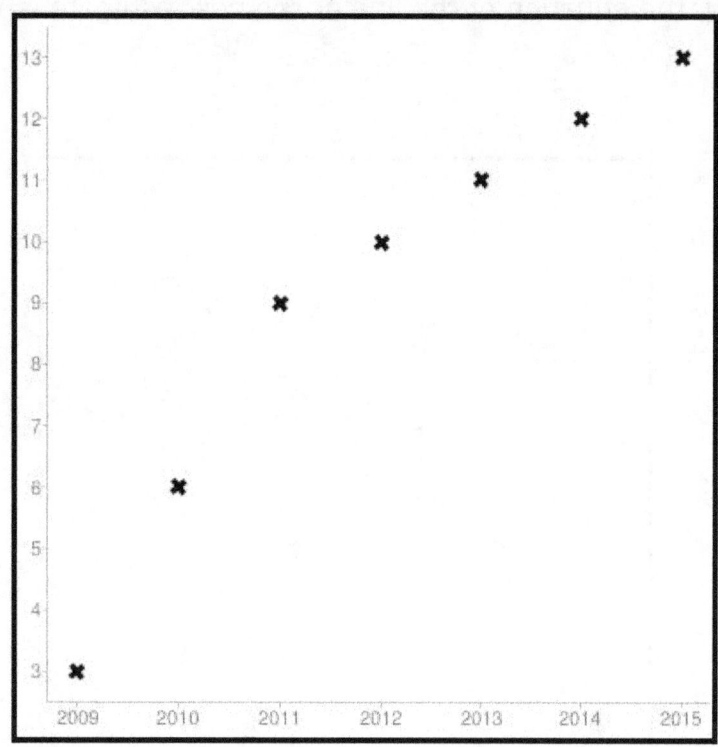

In order to build the linear regression equation and the regression line, let's remember we need the slope and the y-intercept.

THIS IS JUST AN ESTIMATE, since the points do not perfectly fit the line. Since "x" is years, let's make 2009 the value for x=1, 2010 the value for x=2, 2011 the value for x=3, etc.
(1,3),(2,6),(3,9),(4,10),(5,11),(6,12),(7,13)

Pick two points, I will pick (3,9) and (5,11).

Now find the slope:

m=(11-9)/(5-3)= 2/2= 1 m=1, then we can find b.

y=mx+b   I will pick (5,11)

Let's plug it into the equation: 11=(1)(5)+ b   11=5+b   b=6

Finally, the equation of the line or the best estimate is Y= x+6

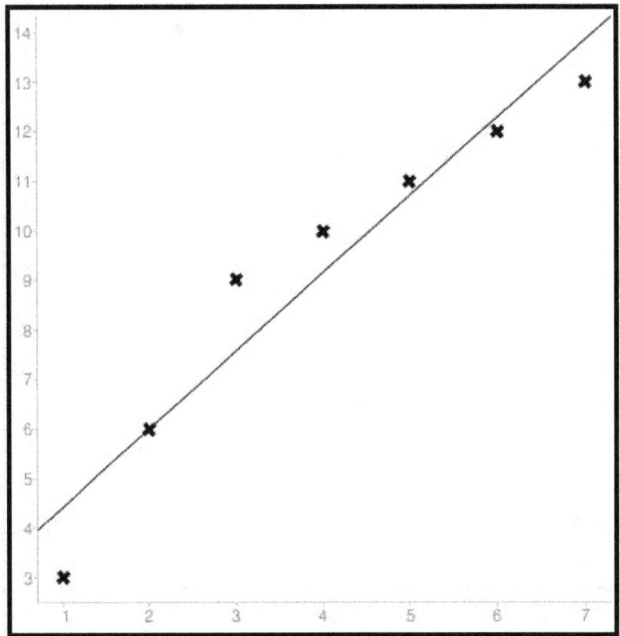

The graph shows a **strong positive correlation** between the increase of the years and the increase of online schools in Florida.

Quick Arithmetic Summary

1. **Integers** can be positive and negative

    $\{-\infty \ldots \ldots, -3, -2, -1, 0, 1, 2, 3 \ldots \ldots \infty\}$

2. **Zero** is an even integer

3. **Adding and subtracting numbers:**

    <u>Same signs</u> you add and keep the sign

    $-3 - 5 = -8; \quad 3 + 5 = 8; \quad -1 - 7 = -8$

    <u>Different signs</u> you subtract and keep sign of the largest number

    $-12 + 2 = -10; \quad 10 - 2 = 8; \quad -5 + 3 = -2$

4. **Absolute value** of a number is the distance from the number to zero on the

    number line. In other words, **the absolute value is <u>always positive</u>**

    $|-3| = 3; \quad |-2 - 3| = |-5| = 5; \quad |-3 + 6| = |3| = 3$

    If there is a negative outside the absolute value, then your result will be negative.

    $-|-3| = -3; \quad -|-2 - 3| = -|-5| = -5$

5. **Factors divisors** are the numbers that evenly divide a number

    The factors of 6 are :   1, 2, 3, 6
    The factors of 18 are :   1, 2, 3, 6, 9, 18

6. **Multiples** are the results of the multiplication of a number with another number

    The multiples of 2 are : $2, 4, 6, 8, 10, 12, \ldots \ldots \ldots \ldots \infty$
    The multiples of 11 are : $11, 22, 33, 44, 55, 66, \ldots \ldots \ldots \ldots \infty$

7. **Divisibility Rules:**

Divisible by 2: The last digit is even
   4, 6, 8, 10, 200, 312

Divisible by 3: Add the digits, the result must be multiple of 3
   51   (5 + 1 = 6 is divisible by 3)
   111   (1 + 1 + 1 = 3 is divisible by 3)
   801   (8 + 0 + 1 = 9 is divisible by 3)

Divisible by 4: The last two digits is a multiple of 4
   832 (32 is a multiple of 4; therefore 832 is divisible by 4)
   9044 (44 is a multiple of 4; therefore 9044 is divisible by 4)

Divisible by 5: The last digit is five or zero
   15, 200, 3035, 1000, 2015

Divisible by 6: The number must be divisible by 2 and 3 at the same time.
   12 ⇨ is divisible by 2 and 3 so it is also divisible by 6
   750 ⇨ is divisible by 2 and 3 so it is also divisible by 6

8. **Prime Number** is an integer greater than 1 with factors 1 and itself.

> **1 IS NOT A PRIME NUMBER**

2, 3, 5, 7, 11, 13, 17, 19, 23, 29, ...
2 is the only <u>even</u> prime number

9. **Composite:** Any integer that is not prime
   4, 6, 8, 15, 200, ...

10. **Prime Factorization:** Is the result of multiplying factors that are Prime, like:
    2, 3, 5, 7, 11

    The prime factorization of 120 = 2*2*2*3*5 = $2^3 * 3 * 5$

11. **Least Common Multiple** (LCM) of two nonzero integers is the least positive **multiple** of both numbers.
    Find the LCM of 200 and 60
    STEP ONE: Find the prime factorization
    $200 = 2^3 * 5^2$
    $60 = 2^2 * 3 * 5$
    STEP TWO: Multiply all the **common** and **non-common** factors with the **highest exponent**.
    $2^3 * 5^2 * 3 =$
    $8 * 25 * 3 = 600$

    Find the LCM of 20 and 30
    $20 = 2^2 * 5$
    $30 = 2 * 5 * 3$

    LCM of 20 and 30 = $2^2 * 5 * 3 = 60$

12. **Greatest Common Factor** (GCF) of two nonzero integers is the greatest
    Positive integer that is a **divisor** of both numbers.
    Find the GCF of 60 and 45

STEP ONE: First do the prime factorization of 60 and 45

$$60 = 2^2 * 3 * 5$$
$$45 = 3^2 * 5$$

STEP TWO: Multiply only the **common** factors with the **lowest exponent**

$$= 3 * 5$$
$$GCF = 15$$

Expressions

Evaluating Expressions

If you need to evaluate an expression, just replace the variable (the letter) with the number given.

Here are some examples:

1) Evaluate: 3A + 3B if A=2 and B=6

Plug the value 2 for every letter A and the value 6 for the letter B

$$3(2) + 3(6) =$$
$$6 + 18 =$$
24

2) Evaluate: 2y-5t-7v if y=1, t=6, v=-1

Plug the value 1 for every letter y, the value 6 for the letter t, and -1 for the letter v.

$$2(1)-5(6)-7(-1)=$$
$$2-30+7=$$
-21

3) Evaluate:

$3a^2 - 2a + 6$ if a=-1

Plug the value of −1 into every letter a:

$$3(-1)^2 - 2(-1) + 6$$
$$3(1)+2+6$$
$$3+2+6$$
11

Let's Practice

1) Evaluate the following expression if a= -1

$$3a^2 - 2a + 5$$

2) Evaluate the following expression if a= -1 and b=2

$$2a - b + 7$$

3) Evaluate the following expression if a= -1 and b=1

$$2 - |3a + b| - 5$$

4) Evaluate the following expression if a= -1 and b=-3

$$\frac{a^2 - b^2}{3a + b}$$

5) Evaluate the following expression if a=3, b=-2 and c=-1

$$2a - (b + c) - c^2$$

6) Evaluate the following expression if a=3, b=2 and c=4

$$\frac{a}{b} - \frac{1}{c}$$

7) Evaluate the following expression if a=-3, b=2

$$\left(\frac{1}{a}\right)^2 + \left(\frac{1}{b}\right)^2$$

8) Evaluate the following expression if a=-1, b=-2

$$-3a - (2a)^3 + b$$

9) Evaluate the following expression if a=-4, b=-2

$$(a-b)(a+b)$$

Let's check your answers:

1) Evaluate the following expression if a= -1

$$3a^2 - 2a + 5$$
$$3(-1)^2 - 2(-1) + 5$$
$$3(1) + 2 + 5$$
$$3 + 2 + 5 = 10$$

2) Evaluate the following expression if a= -1 and b=2

$$2a - b + 7$$
$$2(-1) - 2 + 7$$
$$-2 - 2 + 7$$
$$-4 + 7 = 3$$

3) Evaluate the following expression if a= -1 and b=1

$$2 - |3a + b| - 5$$
$$2 - |3(-1) + 1| - 5$$
$$2 - |-3 + 1| - 5$$
$$2 - |-2| - 5$$
$$2 - 2 - 5$$
$$0 - 5 = -5$$

4) Evaluate the following expression if a= -1 and b=-3

$$\frac{a^2-b^2}{3a+b}$$

$$\frac{(-1)^2-(-3)^2}{3(-1)+(-3)}$$

$$\frac{1-9}{-3-3}$$

$$\frac{-8}{-6}$$

$$\frac{8}{6}=\frac{4}{3}$$

5) Evaluate the following expression if a=3, b=-2 and c=-1

$$2a-(b+c)-c^2$$

$$2(3)-(-2-1)-(-1)^2$$
$$6-(-3)-(1)$$
$$6+3-1$$
$$9-1=8$$

6) Evaluate the following expression if a=3, b=2 and c=4

$$\frac{a}{b}-\frac{1}{c}$$

$$\frac{3}{2}-\frac{1}{4}$$

$$\frac{(2)3}{(2)2}-\frac{1}{4}$$

$$\frac{6}{4}-\frac{1}{4}=\frac{5}{4}$$

7) Evaluate the following expression if a=-3, b=2

$$\left(\frac{1}{a}\right)^2 + \left(\frac{1}{b}\right)^2$$

$$\left(\frac{1}{-3}\right)^2 + \left(\frac{1}{2}\right)^2$$

$$\frac{1}{9} + \frac{1}{4} = \frac{4+9}{36} = \frac{13}{36}$$

8) Evaluate the following expression if a=-1, b=-2

$-3a - (2a)^3 + b$

$-3(-1) - (2(-1))^3 + (-2)$

$3 - (-2)^3 - 2$

$3 - (-8) - 2$

$3 + 8 - 2 = 9$

9) Evaluate the following expression if a=-4, b=-2

$(a - b)(a + b)$

$(-4 - (-2))(-4 - 2)$

$(-4 + 2)(-4 - 2)$

$(-2)(-6) = 12$

Simplifying Expressions

When simplifying expressions, you need to group all the like terms together.

For example:
Simplify the following expression: 3X+ 2Y-4X+6Y
Group together the terms that are alike: 3X-4X + 2Y+6Y= -X+ 8Y

Let's do another example:
$3(x-5y)-(x+2y)$
Let's do the parenthesis first, by distributing the 3 and the negative (-) in front of the parenthesis
3x-15y-x-2y
Now, just group the like terms:
3x-x-15y-2y
2x-17y

Another example:
$5a(2-b)-4b(6+7a)$
Let's distribute the numbers in front of the parenthesis:
10a-5ab-24b-28ab
Now you can group the like terms:
10a-33ab-24b

Let's Practice Simplifying Expressions:
1) Simplify the following expression:
$3x+4y-6x+5y$

2) Simplify the following expression:
$5a-5ab+6a-(2ab-3a)$

3) Simplify the following expression:
$$3x+5(x-10)-12x+10$$

4) Simplify the following expression:
$$(5x^2-7x+10)-(3x^2+6x-10)$$

5) Simplify the following expression:
$$-3(x-4)-5(x+6)-2x$$

6) Simplify the following expression:
$$3y(2-x)-7y(5+6x)$$

Let's check your answers:

1) Simplify the following expression:
$$3x+4y-6x+5y$$
$$3x-6x+4y+5y$$
$$-3x+9y$$

2) Simplify the following expression:
$$5a-5ab+6a-(2ab-3a)$$
$$5a-5ab+6a-2ab+3a$$
$$5a+3a+6a-5ab-2ab$$
$$14a-7ab$$

3) Simplify the following expression:
$$3x+5(x-10)-12x+10$$
$$3x+5x-50-12x+10$$
$$3x+5x-12x-40$$
$$-4x-40$$

4) Simplify the following expression:

$$(5x^2 - 7x + 10) - (3x^2 + 6x - 10)$$

$$5x^2 - 7x + 10 - 3x^2 - 6x + 10$$
$$5x^2 - 3x^2 - 7x - 6x + 10 + 10$$
$$2x^2 - 13x + 20$$

5) Simplify the following expression:
$$-3(x-4) - 5(x+6) - 2x$$
$$-3x + 12 - 5x - 30 - 2x$$
$$-3x - 5x - 2x + 12 - 30$$
$$-10x - 18$$

6) Simplify the following expression:
$$3y(2-x) - 7y(5+6x)$$
$$6y - 3xy - 35y - 42xy$$
$$6y - 35y - 3xy - 42xy$$
$$-29y - 45xy$$

Translating Expressions

Expressions can be translated from English to a math expression or vice versa.

There are key words to use when you are translating an expression.

> The following are the most commonly used words:
> "**Is**", "**Were**", "**Was**" means "**equals**"
> "**Of**" means "**multiply**"

> **Sum** is addition
> **Difference** is subtraction
> **Product** is multiplication
> **Quotient** is division

The following chart shows some other examples:

Symbol	Read As	Example
=	is, was	Alex is 5 years old A= 5
<	less than, is no greater than	X is no greater than 5 X<5
≤	less than or equal, is at most	The price of the item is at most $6 p≤ 6
>	is greater than, is more than	X is more than Y X>Y
≥	is greater than or equal, is at least	X is at least the same value as Y X≥Y
-	Less than	X is less than 5 5-X
+	More than	X is more than 7 X + 7
2	Twice	Maria is twice Dianna's age M= 2D

Let's do some examples

1. 3 less than x is twice the sum of x and 5.

$$x - 3 = 2(x+5)$$

2. X decreased by 7 is three times the difference of x and 2.

$$x - 7 = 3(x-2)$$

3. Four – fifths of the sum of 4 and y plus the product of 9 and c.

$$4/5(4+y) + 9c$$

4. Vanessa is 13 years older than Mahendra.

$$V = M+13$$

5. The length of a sign is 6 inches longer than its width

$$l = 6 + w$$

Let's Practice!

1) The length of a rectangle is four times the width

2) Three times a number is twice the number less than 5

3) Peter is 10 years younger than Ana

4) Five more than twice a number is three times the number

decreased by eleven

Let's check your answers:

1) L=4W

2) 3x=5-2x

3) P=A-10

4) 5+2x=3x-11

Exponential Expressions

Let's learn the rules to treat exponents with examples:

Rule 1: When you multiply exponents with the same base, you need to keep the base and add the exponents on the top:

$$x^3 x^2 = x^{3+2} = x^5$$

Another example:

$$x(x^3 + 3x^2)$$

When you distribute the x, you need to apply the rule of adding the exponents since you are multiplying the base.

$$x^4 + 3x^3$$

Rule 2: When you divide exponents with the same base, you will subtract them

$$\frac{x^5}{x^2} = \frac{x^5}{x^2} = x^{5-2} = x^3$$

Rule 3: When you are raising a power to a power, you need to multiply:

$$(x^3)^6 = x^{3*6} = x^{18}$$

Rule 4: When you have a negative exponent, you can switch the sign by finding the reciprocal or (easier way) just moving it from top to bottom or vice-versa.

$$x^{-3} = \frac{1}{x^3}$$

$$\frac{1}{x^{-5}} = x^5$$

Rule 5: When you raise an exponent to ZERO, the answer is 1.

$$x^0 = 1$$

**Be Careful, if you have a negative in front, then the answer is -1.

$$-x^0 = -1$$

BUT, if you have it with a parenthesis like this, then the answer is 1

$$(-x)^0 = 1$$

Let's do some examples:

1) $(x^3 y^4)^5 = x^{3*5} y^{5*4} = x^{15} y^{20}$

Hint: First distribute the 5 to each base, then simplify

2) $\dfrac{x^{-3} y^6}{x^4 y^{-4}} = \dfrac{y^6 y^4}{x^3 x^4} = \dfrac{y^{10}}{x^7}$

Hint: First move the negative exponents to the opposite side to make them positive, for example, x^{-3} to the bottom and y^{-4} to the top, then you can add the exponents and simplify the rest.

3) $(2x^{-3} y^5)^{-1} = (2^{-1} x^3 y^{-5}) = \dfrac{x^3}{2y^5}$

Hint: The first step is to distribute the -1 to every single base, including the 2, then you can leave the positive base on the top and the negatives on the bottom to switch the exponent

Let's Practice!

1) Simplify the following exponential expression:

$$(3xyz)^3$$

2) Simplify the following exponential expression:

$$(5y^2)^{-3}$$

3) Simplify the following exponential expression:

$$(x^2y^3)^2(x^8y^4)^3$$

4) Simplify the following exponential expression:

$$\frac{x^{-3}}{x^{-5}}$$

5) Simplify the following exponential expression:

$$(4z^2y^{-2})^2(3z^5y^{10})^{-2}$$

6) Simplify the following exponential expression:

$$\left(\frac{2x^5y^{-3}}{3x^{-6}y^5}\right)^{-1}$$

Let's check your answers

1) Simplify the following exponential expression:
$$(3xyz)^3$$
$$3^3 x^3 y^3 z^3$$
$$27 x^3 y^3 z^3$$

2) Simplify the following exponential expression:
$$(5y^2)^{-3}$$
$$5^{-3} y^{2(-3)}$$
$$5^{-3} y^{-6}$$
$$\frac{1}{5^3 y^6} = \frac{1}{125 y^6}$$

3) Simplify the following exponential expression:
$$(x^2 y^3)^2 (x^8 y^4)^3$$
$$x^{2\cdot 2} y^{3\cdot 2} x^{8\cdot 3} y^{4\cdot 3}$$
$$x^4 y^6 x^{24} y^{12}$$
$$x^{4+24} y^{6+12}$$
$$x^{28} y^{18}$$

4) Simplify the following exponential expression:
$$\frac{x^{-3}}{x^{-5}}$$

First switch the exponents from bottom to top and vice versa to

$$\frac{x^5}{x^3}$$

change the sign. x^{5-3}

$$x^2$$

5) Simplify the following exponential expression:

$$(4z^2y^{-2})^2(3z^5y^{10})^{-2}$$

$$4^2z^{2\cdot 2}y^{-2\cdot 2}3^{-2}z^{-2\cdot 5}y^{10\times -2}$$

$$16z^4y^{-4}\frac{1}{9}z^{-10}y^{-20}$$

$$\frac{16}{9}z^{4-10}y^{-4-20}$$

$$\frac{16}{9}z^{-6}y^{-24}$$

$$\frac{16}{9z^6y^{24}}$$

6) Simplify the following exponential expression:

$$\left(\frac{2x^5y^{-3}}{3x^{-6}y^5}\right)^{-1}$$

$$\frac{2^{-1}x^{5\cdot-1}y^{-3\cdot-1}}{3^{-1}x^{-6\cdot-1}y^{5\cdot-1}}$$

$$\frac{2^{-1}x^{-5}y^3}{3^{-1}x^6y^{-5}}$$

$$\frac{3y^5y^3}{2x^6x^5}=\frac{3y^8}{2x^{11}}$$

Multiplying Expressions

When you multiply polynomials, you need to follow the rules of exponents.

Let's do an example:

$$xy(3x - y^2)$$

Distribute the variables xy to every single term:

$$(3xxy - y^2 y)$$
$$3x^2 y - y^3$$

Let's do another example:
Simplify
(3x-y)(x+2y)

STEP ONE: Multiply the 3x by every single term on the second parenthesis
and the same with -y

STEP TWO: Just group the terms

$$3x^2 + 6yx - xy - y^2$$
$$3x^2 + 5xy - y^2$$

Another example:
$$(a - b)(a + b)$$
$$a^2 + ab - ab - b^2 \text{ (Group the terms alike)}$$
$$a^2 + 0 - b^2$$
$$a^2 - b^2$$

Simplify the following polynomials:

1) $(x-4)(x+5)=$

2) $xy(x^3-y^5)=$

3) $(2x-y)(x+3y)=$

4) $(x^2-y^5)(4x^2-y^5)=$

5) $-2y(5y^2+7xy-6)=$

6) $(3x-2y)^2=$

Let's check your answers

1) $(x-4)(x+5) = x^2+5x-4x-20$

$= x^2+x-20$

2) $xy(x^3-y^5) = x^4y-xy^6$

3) $(2x-y)(x+3y) = 2x^2+6xy-xy-3y^2$
$= 2x^2+5xy-3y^2$

4) $(x^2-y^5)(4x^2-y^5) = x^4-x^2y^5-4x^2y^5+y^{10}$
$= x^4-5x^2y^5+y^{10}$

5) $-2y(5y^2+7xy-6) = -10y^3-14xy^2+12y$

6) $(3x-2y)^2 = (3x-2y)(3x-2y) = 9x^2-6xy-6xy+4y^2$
$= 9x^2-12xy+4y^2$

Factoring

Factoring by Greatest Common Factor

Factoring is one of the most important lessons for algebra. It is really important you learn how to factor so you can ACE this test!

Let's start with factoring by the GREATEST COMMON FACTOR:

For example:
$$a^8 - a^5$$

STEP ONE: You need to find the Greatest Common Factor, in this case is the letter with the lowest exponents a^5.

STEP TWO: Now extract that term from both letters and you get:

$a^5(a^3 - 1)$, as you can see, you are just rewriting the binomial.

You can check your answer by just multiplying again.

Let's do another example:

$$4x^2y - 2xy$$

The GCF is 2xy, that is the most you can extract from the binomial
2xy(2x-1)

Last example:
$$6a^2b - 2ab + 12$$
Here the GCF is only 2, since there is not GCF for the letters.

$$2(3a^2b - ab + 6)$$

Factor the following polynomials by the Greatest Common Factor (GCF) method

1) $16x^4y - 6xy$

2) $24x^2 - 12x + 6$

3) $3a^7 - 6a^3 - 2a$

4) $14x^6y^3 - 2x^4y^2 + xy$

Answers: 1) $2xy(8x^3 - 3)$ **2)** $6(4x^2 - 2x + 1)$ **3)** $a(3a^6 - 6a^2 - 2)$

4) $xy(14x^5y - 2x^3y + 1)$

Factoring by Grouping

When you factor by grouping you will have four terms. You need to group the common terms the following way:

$$x^2 + 4x + 5x + 20$$

STEP ONE: Find the GCF of the first two terms ($x^2 + 4x$), in this case x.

STEP TWO: Now find the GCF of the last two terms (5x+20), in this case 5.

STEP THREE: Then you will have two parentheses (they must be equal to each other) $x(x+4) + 5(x+4)$

STEP FOUR: Finally put the two terms outside together (x+5)(x+4)

Let's do another example:
Group by factoring the following polynomial:

$$2x^2 - xy + 2xy - y^2$$

First find the GCF
$$x(2x-y)+y(2x-y)$$
Finally, just put together the factors
(x+y)(2x-y)

Let's Practice!
Factor the following polynomials by grouping

1) Factor $x^2 - xy + 3xy - 3y^2$

2) Factor $2a^2 + 2ab - ab - b^2$

3) Factor $10x^2 - 5xy + 2xy - y^2$

4) Factor $8x^4 + x^2 + 8x^2 + 1$

5) Factor $x^2 - 10x - 3x + 30$

Let's check your answers

1) Factor $x^2 - xy + 3xy - 3y^2$

$x^2 - xy + 3xy - 3y^2$
$x(x-y) + 3y(x-y)$
$(x+3y)(x-y)$

2) Factor $2a^2 + 2ab - ab - b^2$

$2a^2 + 2ab - ab - b^2$
$2a(a+b) - b(a+b)$
$(2a-b)(a+b)$

3) Factor $10x^2 - 5xy + 2xy - y^2$

$10x^2 - 5xy + 2xy - y^2$
$5x(2x-y) + y(2x-y)$
$(5x+y)(2x-y)$

4) Factor $8x^4 + x^2 + 8x^2 + 1$

$8x^4 + x^2 + 8x^2 + 1$
$x^2(8x^2+1) + 1(8x^2+1)$
$(x^2+1)(8x^2+1)$

5) Factor $x^2 - 10x - 3x + 30$

$x^2 - 10x - 3x + 30$
$x(x-10) - 3(x-10)$
$(x-3)(x-10)$

Factoring Trinomials

When you factor trinomials, you need to do the following steps, let's do an example:

Factor the following trinomial
$$x^2 - 2x - 8$$

STEP ONE: Multiply the front coefficient (the value next to x², which is 1) with the last coefficient (which is -8)

1×-8=-8

STEP TWO: Find two factors that when you add/subtract will give you the middle term (which is -2) and when you multiply them give you -8. The only factors that will follow those conditions are -4 and 2

-4×2= -8 and also

-4+2= -2

STEP THREE: Rewrite the trinomial as a polynomial, with the middle terms you just found:
$$x^2 + 2x - 4x - 8$$
As you can see, you just rewrote -2x as 2x-4x

STEP FOUR: Factor by grouping

$$x^2 + 2x - 4x - 8$$
$$x(x+2) - 4(x+2)$$
$$(x-4)(x+2)$$

Let's do another example:
$$2x^2 - x - 6$$

STEP ONE: Multiply the front coefficient (the value next to the x^2) in this case 2 with the last coefficient -6
$$2 \times (-6) = -12$$
STEP TWO: Find two factors that when you add/subtract will give you the middle term (-1) and when you multiply them give you 12. The only factors that will follow for those conditions are -4 and 3
$$-4 \times 3 = -12 \text{ and } -4 + 3 = -1$$
STEP THREE: Rewrite the trinomial with the middle terms you just found as a polynomial:
$$2x^2 - 4x + 3x - 6$$
As you can see, you just rewrote -x as -4x+3x

STEP FOUR: Factor by grouping

$$2x^2 - 4x + 3x - 6$$
$$2x(x-2) + 3(x-2)$$
$$(2x+3)(x-2)$$

Let's Practice!

Factor the following polynomials

1) $x^2 + 2x - 15$

2) $x^2 - 10x + 21$

3) $x^2 - x - 56$

4) $x^2 + 4x - 12$

6) $3x^2 - xy - 4y^2$

7) $4a^2 + 9ab - 9b^2$

8) $35x^2 - 3x - 2$

9) $24x^2 - 13x - 2$

5) $2x^2 + 5x - 3$ 10) $x^2 + 2x + 1$

Let's check your answers:

1) $x^2 + 2x - 15$ = $(x - 3)(x + 5)$

2) $x^2 - 10x + 21$ = $(x - 3)(x - 7)$

3) $x^2 - x - 56$ = $(x + 7)(x - 8)$

4) $x^2 + 4x - 12$ = $(x - 2)(x + 6)$

5) $2x^2 + 5x - 3$ = $2x^2 - x + 6x - 3$ = $x(2x - 1) + 3(2x - 1)$
= $(2x - 1)(x + 3)$

6) $3x^2 - xy - 4y^2$ = $3x^2 + 3xy - 4xy - 4y^2$
= $3x(x + y) - 4y(x + y)$
= $(3x - 4y)(x + y)$

7) $4a^2 + 9ab - 9b^2$ = $4a^2 - 3ab + 12ab - 9b^2$
= $a(4a - 3b) + 3b(4a - 3b)$
= $(a + 3b)(4a - 3b)$

8) $35x^2 - 3x - 2$ = $35x^2 + 7x - 10x - 2$
= $7x(5x + 1) - 2(5x + 1)$
= $(7x - 2)(5x + 1)$

9) $24x^2 - 13x - 2$ = $24x^2 - 16x + 3x - 2$
= $8x(3x - 2) + 1(3x - 2)$

$= (8x + 1)(3x - 2)$

10) $\quad x^2 + 2x + 1 = (x + 1)(x + 1) = (x + 1)^2$

Difference of Squares

When you have a binomial (this means two terms) where the terms are perfect squares (that means something like 4,9,16,25,64, 81..) then you can use the following formula:

> **FORMULA FOR DIFFERENCE OF SQUARES:**
> $a^2 - b^2 = (a - b)(a + b)$

Let's do some examples:
$$x^2 - 1 = (x - 1)(x + 1)$$
$$x^2 - 25 = (x - 5)(x + 5)$$
$$x^2 - 36 = (x - 6)(x + 6)$$
$$64x^2 - 81y^2 = (8x - 9y)(8x + 9y)$$

As you can see, they have a pattern of being perfect squares AND the sign in the middle is negative (-)

If you have something like these examples, but the sign in the middle is a positive (+), they CANNOT be factored.

SPECIAL CASE FORMULA:
$$a^2 + b^2 = PRIME$$

These examples are called **PRIME**.

$$x^2 + 1$$
$$x^2 + 25$$
$$x^2 + 36$$

Let's Practice!

Remember: $a^2 - b^2 = (a - b)(a + b)$

1. $x^2 - 16 =$
2. $x^2 - 25 =$
3. $x^2 - 1 =$
4. $9x^2 - 4x^2 =$
5. $16x^2y^2 - x^2y^2 =$
6. $a^2 - 25 =$
7. $36 - y^2 =$
8. $x^2 + 16 =$
9. $x^2 + 1 =$

10. $4a^2 - b^2 =$

The following binomials are a bit challenge but you can do them, remember to factor by the GCF if necessary first ☺

11. $a^4 - b^4$
12. $4a^4 - 16b^4$
13. $a^8 - b^8$

Let's check your answers

1. $(x+4)(x-4)$
2. $(x+5)(x-5)$
3. $(x+1)(x-1)$
4. $(3x-2y)(3x+2y)$
5. $(4xy-xy)(4xy+xy)$
6. $(a+5)(a-5)$
7. $(6-y)(6+y)$
8. Prime
9. Prime
10. $(2a-b)(2a+b)$
11. $(a^2-b^2)(a^2+b^2) = (a-b)(a+b)(a^2+b^2)$
12. $4(a^4-4b^4) = 4(a^2+2b^2)(a^2-2b^2)$
13. $(a^4)^2-(b^4)^2 = (a^4-b^4)(a^4+b^4) = (a^2-b^2)(a^2+b^2)(a^4+b^4) = (a-b)(a+b)(a^2+b^2)(a^4+b^4)$

Sum and Differences of Cubes

You can also have the same special case with cubes
(when exponent is a 3 instead of a 2)

FORMULAS FOR SUM AND DIFFERENCE OF CUBES:
$$a^3 + b^3 = (a + b)(a^2 - ab + b^2)$$
$$a^3 - b^3 = (a - b)(a^2 + ab + b^2)$$

You can notice it is the same formulas for the sum and difference. The first term follows the same sign and then you alternate the signs in the second terms. The last term is always positive.

Let's do some examples

$$x^3 + 1$$
Notice that $a = x$ and $b = 1$. Now you can use the formula
$$(x + 1)(x^2 - x + 1)$$

$$8x^3 - 27$$
Notice that $a = (2x)$ and $b = 3$ since those are the cubes of $8x^3$ and 27.
$$8x^3 - 27 = (2x - 1)((2x)^2 + (2x * 3) + 3^2) = (2x - 1)(4x^2 + 6x + 9)$$

They are a pain in the neck but practice will help you master them.

Let's Practice!
1. $x^3 - 27 =$
2. $y^3 - 64 =$
3. $a^3 + 125b^3 =$
4. $343x^3 + 512y^3 =$
5. $16x^3 - 250 =$ (hint, try to factor by the GCF first to obtain the cubes)

Let's check your answers

1. $x^3 - 27 = (x - 3)(x^2 + 3x + 9)$
2. $y^3 - 64 = (y - 4)(y^2 + 4y + 16)$
3. $a^3 + 125b^3 = (a + 5)(a^2 - 5a + 25)$
4. $343x^3 + 512y^3 =$

   $Notice\ that\ a =\ 7x, since\ (7x)^3 is\ 343x^3.\ b =\ 8y\ , since\ (8y)^3 is\ 512y^2$

   $(7x+8y)((7x)^2-(7)(8)xy +(8y)^2)= (7x + 8y)(49x^2 - 56xy + 64y^2)$

5. $16x^3 - 250$

$(hint, try\ to\ factor\ by\ the\ GCF\ first\ to\ obtain\ the\ cubes)$

$16x^3 - 250\ (2\ is\ the\ GCF\ between\ 16\ and\ 50)$

$2(8x^3 - 125)\ (finally\ following\ the\ formula\ where\ a =\ 2x\ and\ b = 5$

$2(8\ x^3 - 125) = 2(2x - 5)(2x^2 + 10x + 25)$

Rational Expressions:
Adding and Subtracting

In order to add or subtract a rational expression, you need the **SAME denominator.**

Let's do some examples:
Add the following rational expression:

$$\frac{2}{x} + \frac{3}{2x}$$

STEP ONE: We need the same denominator. Multiply the first term by "2" (top and bottom)

$$\frac{2*2}{2*x} + \frac{3}{2x}$$

$$\frac{4}{2x} + \frac{3}{2x}$$

STEP TWO: Now just add the expression

$$\frac{7}{2x}$$

Let's try some more complicated ones:
Adding / Subtracting Rational Expressions

$$\frac{1}{x+3} + \frac{3}{x+2}$$

STEP ONE: You need the same denominator, therefore multiply both fractions by the opposite denominator

$$\frac{(x+2)1}{(x+2)(x+3)} + \frac{3(x+3)}{(x+2)(x+3)}$$

$$\frac{(x+2)}{(x+2)(x+3)} + \frac{3x+9}{(x+2)(x+3)}$$

STEP TWO: Now that you have the same denominator, you can add the fractions.

$$\frac{x+2+3x+9}{(x+2)(x+3)} = \frac{4x+11}{(x+2)(x+3)}$$

Let's Practice!

1. $\frac{3}{x} + \frac{2}{y}$

2. $\frac{3}{x+2} - \frac{5}{x-1}$

3. $\frac{b}{a} + \frac{3b}{c}$

4. $\frac{x+3}{x+5} + \frac{x-3}{x-5}$

5. $\frac{8}{7x} + \frac{1}{5y}$

Let's check your answers:

1. $\frac{3(y)}{x(y)} + \frac{2(x)}{y(x)} = \frac{3y+2x}{xy}$

2. $\frac{3(x-1)}{(x+2)(x-1)} - \frac{5(x+2)}{(x+2)(x-1)}$

$$\frac{3x-3-5x-10}{(x+2)(x-1)} = \frac{-2x-13}{(x+2)(x-1)}$$

3. $\dfrac{b(c)}{a(c)} + \dfrac{3b(a)}{c(a)} = \dfrac{bc+3ab}{ac}$

4. $\dfrac{(x+3)(x-5)}{(x+5)(x-5)} + \dfrac{(x-3)(x+5)}{(x-5)(x+5)}$

$$\frac{x^2-5x+3x-15+x^2+5x-3x-15}{(x+5)(x-5)}$$

$$\frac{2x^2-30}{(x+5)(x-5)}$$

5. $\dfrac{8}{7x} + \dfrac{1}{5y}$

$\dfrac{8(5y)}{7x(5y)} + \dfrac{1(7x)}{7x(5y)} = \dfrac{40y+7x}{7x(5y)}$

Rational Expressions:
Multiplying and Dividing

When you multiply rational expressions, you follow the same procedure as multiplying or dividing fractions.
You **DO NOT need the same denominator**.

Let's look at two examples:

$$\frac{x}{3y} \cdot \frac{15x^3}{y^2} = \frac{15x^4}{3y^3} = \frac{5x^4}{y^3}$$

In the example below, you need to factor all the terms of each fraction so that you can simplify the fractions and finally multiply them.

$$\frac{x+3}{x^2+6x+9} \cdot \frac{x^2+x-6}{x^2-9} = \frac{x+3}{(x+3)(x+3)} \cdot \frac{(x+3)(x-2)}{(x+3)(x-3)} = \frac{x-2}{(x+3)(x-3)}$$

Let's Practice!

1. $\dfrac{2x^2-7x-15}{x^2+12x+35} \cdot \dfrac{x^2+5x-14}{2x^2+13x+15}$

2. $\dfrac{a^2-b^2}{a+b} \cdot \dfrac{2}{4a^2-16b^2}$

3. $\dfrac{x+3}{x^2-3x-18} \div \dfrac{2x^2-5x-12}{2x^2-9x-18}$

4. $\dfrac{x+y}{3x+6y} \div \dfrac{2}{x^2+xy-2y^2}$

Let's check your answers

1. $\dfrac{2x^2-7x-15}{x^2+12x+35} \cdot \dfrac{x^2+5x-14}{2x^2+13x+15} = \dfrac{(2x+3)(x-5)}{(x+7)(x+5)} \cdot \dfrac{(x+7)(x-2)}{(x+5)(2x+3)} =$

$\dfrac{(x-5)(x-2)}{(x+5)(x+5)} = \dfrac{(x-5)(x-2)}{(x+5)^2}$

2. $\dfrac{a^2-b^2}{a+b} \cdot \dfrac{2}{4a^2-16b^2} = \dfrac{(a+b)(a-b)}{a+b} \cdot \dfrac{2}{4(a^2-4b^2)}$

$= \dfrac{(a+b)(a-b)}{a+b} \cdot \dfrac{2}{4(a+2b)(a-2b)} = \dfrac{1(a-b)}{2(a+2b)(a-2b)}$

3. $\dfrac{x+3}{x^2-3x-18} \div \dfrac{2x^2-5x-12}{2x^2-9x-18} = \dfrac{x+3}{(x-6)(x+3)} \cdot \dfrac{(2x+3)(x-6)}{(2x+3)(x-4)} =$

$\dfrac{1}{x-4}$ (do not forget you have a 1 on the top)

4. $\dfrac{x+y}{3x+6y} \div \dfrac{2}{x^2+xy-2y^2} = \dfrac{x+y}{3(x+2y)} \cdot \dfrac{x^2+xy-2y^2}{2} = \dfrac{x+y}{3(x+2y)} \cdot$

$\dfrac{(x+2y)(x-y)}{2} = \dfrac{(x+y)(x-y)}{6}$

Complex Fractions

Complex fractions are just two or more fractions to simplify. In the lesson before you learned about rational expressions. Complex fractions are very similar to these expressions.

Let's do an example:

$$\frac{\frac{x-7}{4}}{\frac{3x-1}{8}}$$

Let's rewrite this expression:

$$\frac{\frac{x-7}{4}}{\frac{3x-1}{8}} = \frac{x-7}{4} \div \frac{3x-1}{8}$$

You can now flip the second fraction and multiply it $\frac{x-7}{4}$ *

$$\frac{8}{3x-1}$$

$$\frac{8(x-7)}{4(3x-1)} = \frac{2(x-7)}{(3x-1)} =$$

Let's Practice!

1. $\dfrac{\frac{1}{4}}{\frac{3}{8}}$

2. $\dfrac{\frac{x}{a} - \frac{y}{a}}{\frac{x+y}{a}}$

3. $\dfrac{\frac{4}{3x}}{\frac{6}{4x}-2}$

4. $\dfrac{\frac{1}{5}-x}{\frac{2}{7}-x}$

5. $\dfrac{\frac{1}{x+y}-2}{3+\frac{2}{x+y}}$

Let's check your answers

1. $\dfrac{\frac{1}{4}}{\frac{3}{8}} = \dfrac{\frac{1(2)}{4(2)}}{\frac{3}{8}} = \dfrac{\frac{2}{8}}{\frac{3}{8}} = \dfrac{2}{8} \div \dfrac{3}{8} = \dfrac{2}{8} \times \dfrac{8}{3} = \dfrac{2}{3}$

2. $\dfrac{\frac{x}{a}-\frac{y}{a}}{\frac{x+y}{a}} = \dfrac{\frac{x-y}{a}}{\frac{x+y}{a}} = \dfrac{x-y}{a} \div \dfrac{x+y}{a} = \dfrac{x-y}{a} \times \dfrac{a}{x+y} = \dfrac{x-y}{x+y}$

3. $\dfrac{\frac{4}{3x}}{\frac{6}{4x}-2} = \dfrac{\frac{4}{3x}}{\frac{6}{4x}-\frac{2(4x)}{1(4x)}} = \dfrac{\frac{4}{3x}}{\frac{6-8x}{4x}} = \dfrac{4}{3x} \div \left(\dfrac{6-8x}{4x}\right)$

$= \dfrac{4}{3x} \times \dfrac{4x}{6-8x} = \dfrac{16}{3(6-8x)} = \dfrac{16}{18-24x} = \dfrac{8}{9-12x}$

4. $\dfrac{\frac{1}{5}-\frac{x}{1}}{\frac{2}{7}-\frac{x}{1}} = \dfrac{\frac{1}{5}-\frac{x(5)}{1(5)}}{\frac{2}{7}-\frac{x(7)}{1(7)}} = \dfrac{\frac{1-5x}{5}}{\frac{2-7x}{7}} = \dfrac{1-5x}{5} \div \dfrac{2-7x}{7}$

$$= \frac{1-5x}{5} \cdot \frac{7}{2-7x} = \frac{7(1-5x)}{5(2-7x)}$$

5. $\dfrac{\frac{1}{x+y} - 2}{\frac{3}{1} + \frac{2}{x+y}} = \dfrac{\frac{1}{x+y} - \frac{2(x+y)}{x+y}}{\frac{3(x+y)}{1(x+y)} + \frac{2}{x+y}} = \dfrac{\frac{1-2x-2y}{x+y}}{\frac{3x+3y+2}{x+y}}$

$$= \frac{1-2x-2y}{x+y} \div \frac{3x+3y+2}{x+y} = \frac{1-2x-2y}{x+y} \times \frac{x+y}{3x+3y+2}$$

$$= \frac{1-2x-2y}{3x+3y+2}$$

Equations

Linear Equations

Linear equations are defined by a function with only one answer.
The term "solving" means to leave "x" alone

For example:
Solve the following linear equation
$$3(x - 4) = 5x + 6$$

STEP ONE: Take care of the parentheses by distributing the 3.
$$3x - 12 = 5x + 6$$

STEP TWO: Group the variables and the numbers together
$$3x - 5x = 12 + 6$$

STEP THREE: Leave "x" alone
$$-2x = 18$$
$$x = \left(-\frac{18}{2}\right) = -9$$
$$x = -9$$

Final Answer $x = -9$

Linear Equations with Fractions

Let's say you have an equation like this one:

$$4x + \frac{1}{2} = x - 6$$

In order to solve for "x" you want to eliminate the denominator $\left(\frac{1}{2}\right)$.

The best way is to multiply every term by 2 because $2\left(\frac{1}{2}\right) = \frac{2}{2} = 1$.

Let's do it!

STEP ONE: Multiply each term by 2 (the denominator of the fraction)

$$(2)4x + (2)\frac{1}{2} = (2)x - (2)6$$

You get: $8x + 1 = 2x - 12$

STEP TWO: Now just solve for x.

$$8x - 2x = -12 - 1$$
$$6x = -13$$
$$x = \frac{-13}{6}$$

Let's do another example:

$$\frac{x}{4} + \frac{3}{2} = 3x - 4$$

In this case we have two fractions, therefore let's multiply by the Least Common Multiple or by a number that will eliminate the 4 and the 2 in the bottom. The Least Common Multiple in this case is "4".

Let's do it!

$$4\left(\frac{x}{4}\right) + 4\left(\frac{3}{2}\right) = 4(3x) - 4(4)$$

$$x + 6 = 12x - 16$$

$$x - 12x = 16 - 6$$

$$-11x = -22$$

$$x = \left(\frac{-22}{-11}\right) = 2$$

Let's Practice!

Solve the following equations

1. $5(x + 3) = 2(x - 2);'$

2. $-3(x + 5) - 2 = 5x + 12$

3. $\frac{3}{x+5} = \frac{2}{x+10}$

4. $6x - 2(3x - 5) = 2x - 5x$

5. $\frac{1}{2}(x + 6) - 3 = 6x - 5$

6. $\frac{3x}{5} - \frac{x}{2} + 6 = 0$

7. $\frac{6x}{5} = \frac{4}{7}$

8. $\frac{x}{0.5} = 3$

Let's check your answers

1. $5(x + 3) = 2(x - 2)$
   $5x + 15 = 2x - 4$
   $5x - 2x = -15 - 4$
   $3x = -19$
   $x = -\frac{19}{3}$

2. $-3(x + 5) - 2 = 5x + 12$
   $-3x - 15 - 2 = 5x + 12$
   $-3x - 17 = 5x + 12$
   $-3x - 5x = 12 + 17$
   $-8x = 29$
   $x = -\frac{29}{8}$

3. $\frac{3}{x+5} = \frac{2}{x+10}$
   $3(x + 10) = 2(x + 5)$
   $3x + 30 = 2x + 10$
   $3x - 2x = 10 - 30$
   $x = -20$

4. $6x - 2(3x - 5) = 2x - 5x$
   $6x - 6x + 10 = -3x$
   $10 = -3$
   $x = -\frac{10}{3}$

5. $\frac{1}{2}(x + 6) - 3 = 6x - 5$

   Multiply everything by 2

$$2 * \frac{1}{2}(x+6) - 2 * 3 = 2 * 6x - 2 * 5$$

$$(x+6) - 6 = 12x - 10$$

$$x + 6 - 6 = 12x - 10$$

$$x = 12x - 10$$

$$-11x = -10$$

$$x = \frac{-10}{-11}$$

$$x = \frac{+10}{11}$$

6. $\quad \frac{3x}{5} - \frac{x}{2} + 6 = 0$

$$\frac{3x}{5} - \frac{x}{2} = -6$$

Multiply by 10 to eliminate the denominators

$$10\left(\frac{3x}{5}\right) - 10\left(\frac{x}{2}\right) = 10(-6)$$

$$2(3x) - 5(x) = -60$$

$$6x - 5x = -60$$

$$x = -60$$

7. $\quad \frac{6x}{5} = \frac{4}{7}$

Multiply by 35 to eliminate the 5 and 7

$$35\left(\frac{6}{5}\right)x = \frac{4}{7}(35)$$

$$7(6x) = 4(5)$$

$$42x = 20$$

$$x = \frac{20}{42} = \frac{10}{21}$$

8. $\frac{x}{0.5} = 3$

Multiply by 0.5 both sides to eliminate it

$0.5\left(\frac{x}{0.5}\right) = 0.5(3) = 1.5$

Literal Equations

These equations are called "literal" because they have different letters. The logic of solving them is the same as the linear equations.

Let's do an example
Solve for "w"
$p = 2l + 2w$

STEP ONE: You need to move all the terms that do not have a "w" to the other side of the equation:

$p - 2l = 2w$

STEP TWO: Finally, we divide by 2 both sides to eliminate the denominator and leave "w" by itself

$\frac{p-2l}{2} = w$

$w = \frac{p-2l}{2}$

Let's Practice!

1) Solve for the height h.

$$A = \frac{b.h}{2}$$

2) Solve for b_1.

$$A = \frac{(b_1+b_2).h}{2}$$

3) Solve for the Radius

$$C = 2\pi R$$

4) Solve for the radius

$$A = \pi * R^2$$

5) Solve for V

$$K_e = \frac{1}{2}m.V^2$$

6) Solve for x

$$ax - y = 6x$$

Let's check your answers

1) Solve for the height

$$A = \frac{b.h}{2}$$

$$2A = \frac{b.h}{2} . 2$$

$$2A = b.h$$

$$h = \frac{2A}{b}$$

2. $A = \frac{(b_1+b_2) \cdot h}{2}$ Solve for b_1

$$2A = (b_1 + b_2)h$$

$$\frac{2A}{h} = b_1 + b_2$$

$$b_1 = \frac{2A}{h} - b_2$$

3. Solve for the Radius
$$C = 2\pi R$$
$$\frac{C}{2\pi} = R \quad \text{or} \quad R = \frac{C}{2\pi}$$

4. Solve for the Radius $A = \pi R^2$

$$A = \pi R^2$$

$$\frac{A}{\pi} = R^2$$

$$\sqrt{\frac{A}{\pi}} = \sqrt{R^2}$$

Remember that when you have a square, you need to square root both sides to eliminate the exponent.

$$R = \sqrt{\frac{A}{\pi}}$$

5. Solve for V

$$K_e = \frac{1}{2} \cdot m \cdot V^2$$

$$2K_e = m \cdot V^2$$

$$\frac{2K_e}{m} = V^2$$

$$\sqrt{\frac{2K_e}{m}} = \sqrt{V^2}$$

Remember that when you have a square, you need to square root both sides to eliminate the exponent.

$$V = \sqrt{\frac{2K_e}{m}}$$

6. Solve for x

$$ax - y = 6x$$
$$ax - 6x = y$$
$$x(a - 6) = y$$
$$x = \frac{y}{a-6}$$

Inequalities

Inequalities are used when you have a range of solutions, for example if I say x=4, that is an equation because x is equal to 4. An inequality such as x>4 means that x is greater than 4, it could be 4.01, 4.05, 5, 1000 and so on.

Let's remember

$\geq$ or $\leq$; This means you are including the number as part of your answer. For example, $x \geq 9$. X can be greater or equal to 9. Your answer will need to include "[" or "] " to set your interval notation answer.

$>$ or $<$; This means you are NOT including the number as part of your answer. For example, $x > 9$, x has be greater than 9. You need to use "(" or ")" to set your interval notation answer.

To solve an inequality, it is the same procedure as an equation.

You just need to make sure you understand how to graph your solution.

Let's do an example:

$$2x - 5 \leq 6 + 3x - 2$$
$$2x - 3x \leq 6 - 2 + 5$$
$$-x \leq 9$$

> WHEN YOU DIVIDE A NEGATIVE NUMBER IN AN INEQUALITY EQUATION, YOU NEED TO SWITCH THE INEQUALITY SYMBOL THE OPPOSITE DIRECTION

(since you are dividing by a negative number (in this case -1) you need to switch the sign (≤ to ≥) and the change the sign of 9 to -9

$$x \geq -9$$

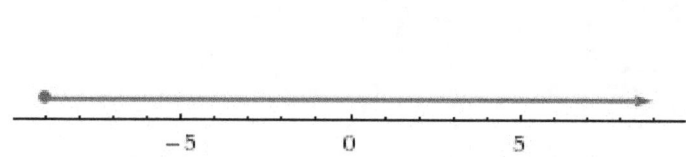

That is, x needs to be a number equal to or greater than -9:

-9,-8,-7,-1....8... ∞

The interval notation solution is:
$$[-9, \infty)$$

In order to graph the solution of an inequality think of a train that goes from $-\infty$ to ∞. This train travels from left to right; in this case it starts on -9 and goes to infinity.

> **THE INFINITY SIGNS ALWAYS HAVE PARENTHESIS**

Let's practice!!!

1. $x + 5 < -3$
2. $3(x + 2) \geq -6$
3. $-5x - 2(x - 3) \geq -1 + 2x$
4. $7x + 2 > -3(x + 5) - 3$
5. $0 \leq 2x - 6$
6. $\frac{-x}{5} \leq 10$
7. $\frac{x}{2} - 2 \geq \frac{x}{3} + 5$

Let's check your answers:

1. $x + 5 < -3$
   $x < -3 - 5$
   $x < -8$
   $(-\infty, -8)$

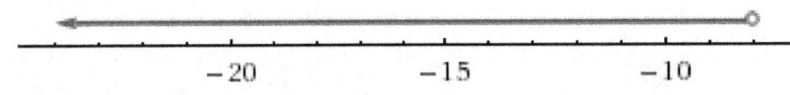

2. $3(x+2) \geq -6$

   $3x + 6 \geq -6$

   $3x \geq -6 - 6$

   $3x \geq -12$

   $x \geq \frac{-12}{3}$

   $x \geq -4$

   $[-4, \infty)$

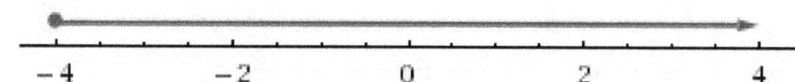

3. $-5x - 2(x-3) \geq -1 + 2x$

   $-5x - 2x + 6 \geq -1 + 2x$

   $-7x + 6 \geq -1 + 2x$

   $-7x - 2x \geq -1 - 6$

   $-9x \geq -7$

   $x \leq \frac{-7}{-9}$

   (Sign must be flipped since you dividing by a negative number)

   $x \leq \frac{7}{9}$

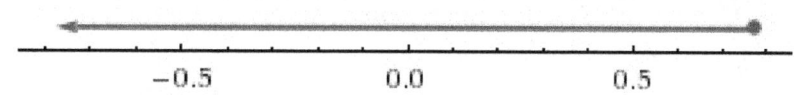

4. $7x + 2 > -3(x + 5) - 3$

$7x + 2 > -3x - 15 - 3$

$7x + 3x > -15 - 3 - 2$

$10x > -20$

$x > \frac{-20}{10}$

$x > -2$

$(-2, \infty)$

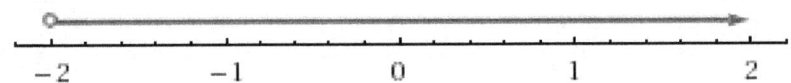

5. $0 \leq 2x - 6 \quad or \quad 2x - 6 \geq 0$

$2x \geq 6$

$x \geq \frac{6}{2}$

$x \geq 3$

$[3, \infty)$

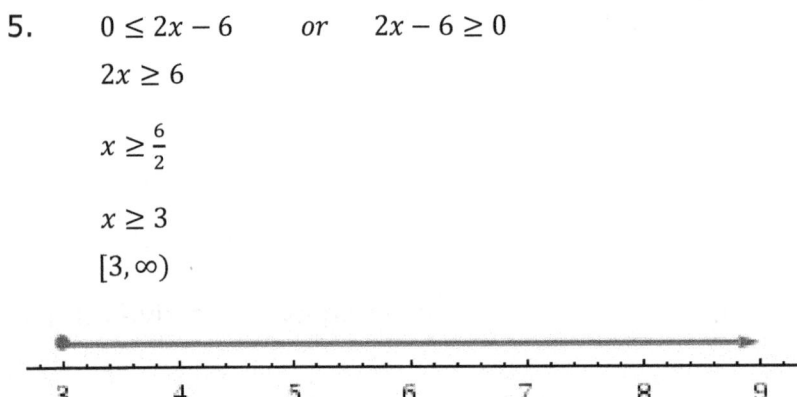

6. $-\frac{x}{5} \leq 10 \;;\; -x \leq 10(5) \;;\; -x \leq 50$

$x \geq -50 \;;$

(You need to flip the sign since you are dividing the inequality by -1

$[-50, \infty)$

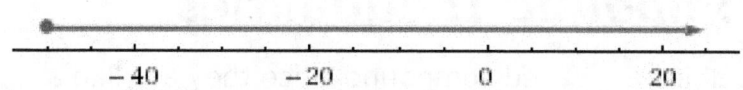

7. $\quad \frac{x}{2} - 2 \geq \frac{x}{3} + 5$

$6\left(\frac{x}{2}\right) - 6(2) \geq 6\left(\frac{x}{3}\right) + 6(5)$

$3x - 12 \geq 2x + 30$

$3x - 2x \geq 30 + 12$

$x \geq 42$

$[42, \infty)$

Compound Inequalities

These inequalities are called compound since they are like a sandwich. You can read them like this:

" x is a number between -9 and 1 but not including -9 and 1 "

$$-9 < x < 1$$

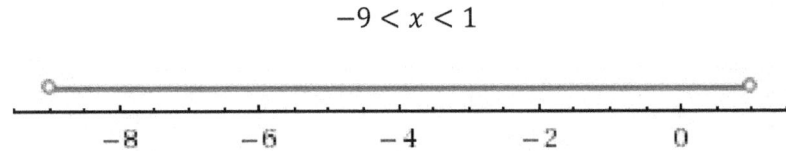

Solve the following compound inequality:

$$10 \leq 6x - 2 \leq 16$$

STEP ONE: Solve both sides of the inequality

$$10 + 2 \leq 6x \leq 16 + 2$$
$$12 \leq 6x \leq 18$$
$$\left(\frac{12}{6}\right) \leq x \leq \left(\frac{18}{6}\right)$$
$$2 \leq x \leq 3$$
$$[2,3]$$

You can read it as: X is between 2 and 3, x can be 2.1, 2.2, 2.3,....,2.99, 3

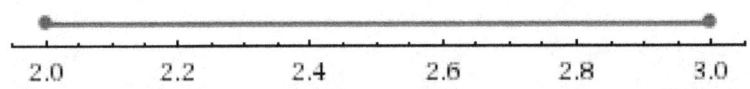

Let's Practice!

1. $3 \leq 2x - 2 \leq 10$
2. $5 < 3 + 2(5x - 4) < 6$
3. $\frac{1}{2} < x - 3 < \frac{7}{2}$ (Hint: multiply every term by 2 to eliminate the denominator)
4. $3 \leq -x \leq 4$ (Hint: Divide Both sides by -1, remember the signs will change)
5. $6 \leq x + 3x - 2 \leq 10$

Let's check your answers

1. $3 \leq 2x - 2 \leq 10$
   $3 + 2 \leq 2x \leq 10 + 2$
   $5 \leq 2x \leq 12$
   $\frac{5}{2} \leq x \leq 6$

   $\left[\frac{5}{2}, 6\right]$

2. $5 < 3 + 2(5x - 4) < 6$
   $5 < 3 + 10x - 8 < 6$
   $5 < 10x - 5 < 6$
   $5 + 5 < 10x < 5 + 6$
   $10 < 10x < 11$
   $1 < x < \frac{11}{10}$

   $\left(1, \frac{11}{10}\right)$

3. $\frac{1}{2} < x - 3 < \frac{7}{2}$

   $2\left(\frac{1}{2}\right) < 2x - 2(3) < 2\left(\frac{7}{2}\right)$

$1 < 2x - 6 < 7$

$1 + 6 < 2x < 7 + 6$

$7 < 2x < 13$

$\frac{7}{2} < x < \frac{13}{2}$

$\left(\frac{7}{2}, \frac{13}{2}\right)$

4. $3 \leq -x \leq 4$

   STEP ONE: Multiply every term by -1 so that $-x$ becomes positive.

   $-3 \geq x \geq -4$

   But remember your signs flip too. Then you can flip everything like a "pancake" to get the final answer.

   $-4 \leq x \leq -3$

   $[-4, -3]$

5. $6 \leq x + 3x - 2 \leq 10$

   $6 \leq 4x - 2 \leq 10$

   $6 + 2 \leq 4x \leq 10 + 2$

   $8 \leq 4x \leq 12$

   $\frac{8}{4} \leq x \leq \frac{12}{4}$

   $2 \leq x \leq 3$

   or $[2, 3]$

Absolute Value Equations

Absolute value equations need to be set up with positive and negative results. I like to think they have a good twin and a bad twin. This means you need to set up the equation with a positive answer (good twin) and another one with a negative answer (bad twin).

For example:
$|3x + 6| = 15$

STEP ONE: We set up the positive answer (good twin). You NEVER change the signs inside the absolute value.

$$3x + 6 = 15$$
$$3x = 15 - 6$$
$$3x = 9$$
$$x = \frac{9}{3} = 3$$

STEP TWO: We set up the negative answer (bad twin).

$$3x + 6 = -15$$
$$3x = -15 - 6$$
$$3x = -21$$
$$x = -\frac{21}{3}$$
$$x = -7$$

The solution is $x = 3$ or $x = -7$

Let's practice!

1) $|2x + 5| = 6$
2) $|2x + 6| - 3 = 9$
3) $-3|7x| = -6$

4) $3 - |5x+6| = 1$
5) $6|8x + 1| = 24$
6) $|x - 3| = -5$

Let's check your answer

1. $|2x + 5| = 6$

    $2x + 5 = 6$ $2x + 5 = -6$

    $2x = 6 - 5$ $2x = -6 - 5$

    $2x = 1$ $2x = -11$

    $x = \frac{1}{2}$ $x = \frac{-11}{2}$

2. $|2x + 6| - 3 = 9$

    $|2x + 6| = 12$

    $2x + 6 = 12$ $2x + 6 = -12$

    $2x = 12 - 6$ $2x = -12 - 6$

    $2x = 6$ $2x = -18$

    $x = \frac{6}{2}$ $x = \frac{-18}{2}$

    $x = 3$ $x = -9$

3. $-3|7x| = -6$

    $|7x| = \frac{-6}{-3}$

    $|7x| = 2$

    $7x = 2$ $7x = -2$

    $x = \frac{2}{7}$ $x = \frac{-2}{7}$

4. $3 - |5x + 6| = 1$
    $-|5x + 6| = 1 - 3$
    $-|5x + 6| = -2$
    $|5x + 6| = -2(-1)$
    $|5x + 6| = 2$

$5x + 6 = 2$	$5x + 6 = -2$
$5x = -6 + 2$	$5x = -6 - 2$
$5x = -4$	$5x = -8$
$x = \frac{-4}{5}$	$x = \frac{-8}{5}$

5. $6|8x + 1| = 24$
    $|8x + 1| = \frac{24}{6}$
    $|8x + 1| = 4$

$8x + 1 = 4$	$8x + 1 = -4$
$8x = 4 - 1$	$8x = -4 - 1$
$8x = 3$	$8x = -5$
$x = \frac{3}{8}$	$x = \frac{-5}{8}$

6) No solution. Since there is no a value for X that will provide a negative answer.

Absolute Value Inequalities

In order to solve absolute value inequalities, you need to follow the following steps:

$$|3x - 6| \geq 15$$

STEP ONE: You need to set up the inequality the following way

$$3x - 6 \geq 15$$
$$3x \geq 15 + 6$$
$$3x \geq 21$$
$$x \geq 7$$

STEP TWO: Now set up the negative inequality (notice the sign of the inequality is flipped and sign of the number is negative)

$$3x - 6 \leq -15$$
$$3x \leq -15 + 6$$
$$3x \leq -9$$
$$x \leq -3$$

Final answer: $x \geq 7$ or $x \leq -3$

Also written as: $(-\infty, -3]$ or $[7, \infty)$

Let's practice!

1. $|x - 2| > 5$
2. $3|x + 6| - 2 \leq 16$
3. $-2|2x + 5| > -6$
4. $3 - |6x - 5| + 1 < 5$
5. $|x - 6| - 5 \geq 6$

Let's check your answers

1. $|x - 2| > 5$

       $x - 2 > 5$ $x - 2 < -5$

       $x > 5 + 2$ $x < -5 + 2$

       $x > 7$ $x < -3$

       $(-\infty, -3) \cup (7, \infty)$

2. $3|x + 6| - 2 \leq 16$

       $3|x + 6| \leq 16 + 2$

       $3|x + 6| \leq 18$

       $|x + 6| \leq \frac{18}{3}$

       $|x + 6| \leq 6$

       $x + 6 \leq 6$ $x + 6 \geq -6$

       $x \leq 6 - 6$ $x \geq -6 - 6$

       $x \leq 0$ $x \geq -12$

       $[-12, 0]$

3. $-2|2x + 5| > -6$

       $|2x + 5| < \frac{-6}{-2}$ (Flip the sign, you are dividing by −2)

       $|2x + 5| < 3$

       $2x + 5 < 3$ $2x + 5 > -3$

       $2x < 3 - 5$ $2x > -3 - 5$

       $2x < -2$ $2x > -8$

       $x < \frac{-2}{-2}$ $x > \frac{-8}{2}$

       $x < -1$ $x > -4$

$(-4, -1)$

4. $\quad 3 - |6x - 5| + 1 < 5$
$4 - |6x - 5| < 5$
$-|6x - 5| < 5 - 4$
$-|6x - 5| < 1$
$|6x - 5| > -1$

$6x - 5 > -1$	$6x - 5 < 1$
$6x > -1 + 5$	$6x < 1 + 5$
$6x > 4$	$6x < 6$
$x > \frac{4}{6}$	$x < \frac{6}{6}$
$x > \frac{2}{3}$	$x < 1$

$\left(\frac{2}{3}, 1\right)$

5. $\quad |x - 6| - 5 \geq 6$
$|x - 6| \geq 6 + 5$
$|x - 6| \geq 11$

$x - 6 \geq 11$	$x - 6 \leq -11$
$x \geq 11 + 6$	$x \leq -11 + 6$
$x \geq 17$	$x \leq -5$

$(-\infty, -5] \cup [17, \infty)$

Coordinate Geometry

Quadrants

The Rectangular Coordinate System has 4 quadrants. Depending where the coordinates (X,Y) are; X and Y can be positive and/or negative.
Look at the graph for examples.

The origin is where X and Y are zero, therefore the origin O has a coordinate of (0,0)

Quadrant
II   Y   I   Origin (0,0)   ——————————— X   III   IV

If you need to plot points, the figure below shows example of coordinates and what Quadrant they will fall into.

The X runs from left to right and Y rises from bottom to top.

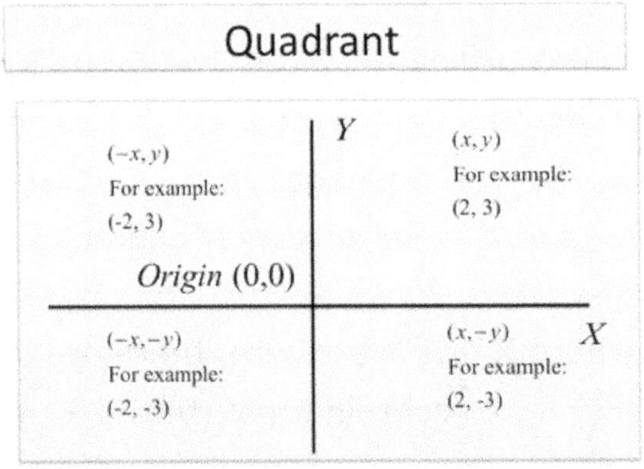

Quadrant

$(-x, y)$	(x, y)
For example:	For example:
(-2, 3)	(2, 3)

Origin (0,0)

$(-x, -y)$	$(x, -y)$
For example:	For example:
(-2, -3)	(2, -3)

Let's practice!

Identify which Quadrant the following coordinates are located:

1. (3,4)
2. (-2,10)
3. (-3,-1)
4. (1,-5)
5. (-3,-3)

Answers: 1) I, 2) II 3) III 4) IV 5) III

Distance and Midpoint

The distance of two points can be found with this formula

$$d = \sqrt{(Y_2 - Y_1)^2 + (X_2 - X_1)^2}$$

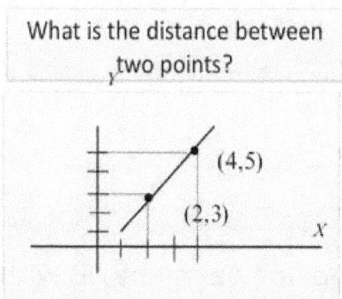

What is the distance between two points?

If you want to find the distance between (2,3) and (4,5) you plug the values into the formula.

Distance between (2,3) and (4,5)

$distance$ between (2,3) and (4,5)
$d = \sqrt{(y_2 - y_1)^2 + (x_2 - x_1)^2}$
$d = \sqrt{(5-3)^2 + (4-2)^2}$
$d = \sqrt{2^2 + 2^2} = \sqrt{4+4} = \sqrt{8}$
$d = 2\sqrt{2}$

The midpoint of two coordinates is found using the following formula

$$X_m = \frac{X_1 + X_2}{2} \quad Y_m = \frac{Y_1 + Y_2}{2}$$

For example:

Find the midpoint between (4,5) and (2,3)

$$X_m = \frac{4+2}{2} = \frac{6}{2} = 3$$

$$Y_m = \frac{5+3}{2} = \frac{8}{2} = 4$$

The midpoint is (3,4)

Let's Practice:

1) Find the distance of the line segment AB with the following coordinates: A(3,5) and B(0,1)

2) Find the distance of line segment AB with the following coordinates: A(-2,3) and B(-3,-4)

3) Find the midpoint of coordinates A(3,6) and B(-5,10)

4) The endpoint of segment AB has coordinates A(1,4) and midpoint (4,9). Find the coordinates of B

Let's check your answers:

1) 5   2) 5√2   3) (-1,8) 4) B(7,14)

Slope of the Line

The slope of the line is just the inclination of the line.
The slope can be positive, negative, zero (no slope), or undefined slope.

Let's see some examples:

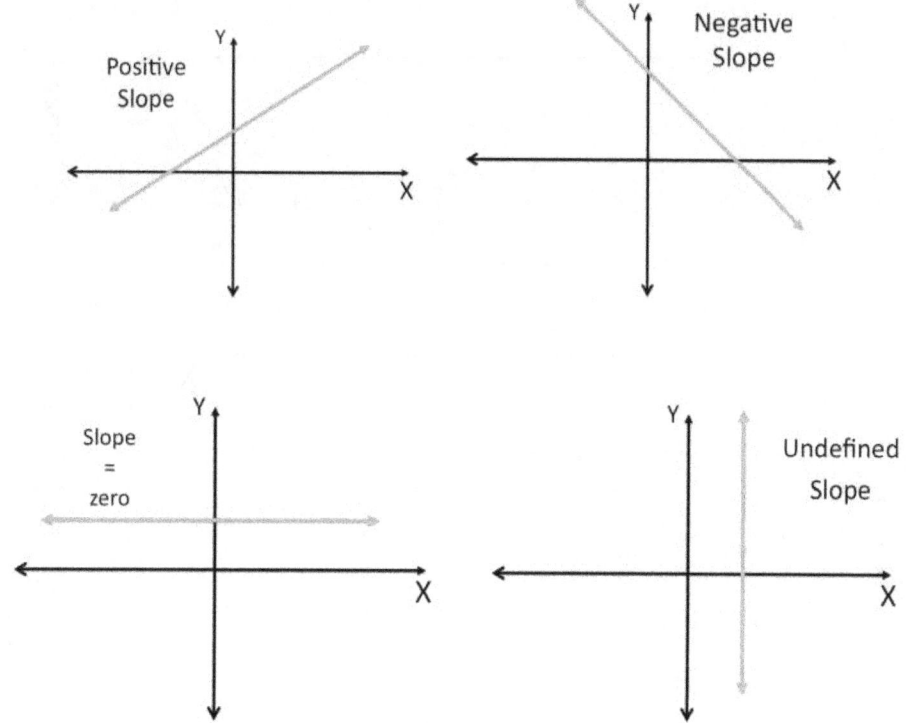

TIPS: Notice that the **positive slope** looks like stairs going up from Quadrant III to the Quadrant I, while the **negative slope** looks like a slide going down from the Quadrant II to the Quadrant IV.
The **zero slope** (looks a like roller coaster for kids) there is no inclination. That means "zero" fun. The **undefined slope** you do not know if it is positive or negative. It is a vertical line.

If you have a graph and need to find the slope, you can to use the Rise over Run Method. This method is very easy, you need to find two points where you can draw a triangle as shown in the figure. You rise how many values the Y values go up and then Run with the X values.

Rise Over Run Method

$$m = \frac{Rise \ (The \ change \ of \ Y)}{Run \ (The \ change \ of \ X)}$$

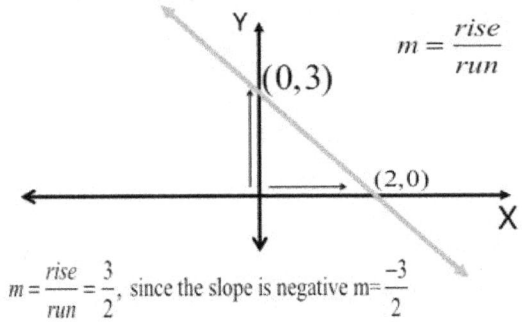

$m = \dfrac{rise}{run} = \dfrac{3}{2}$, since the slope is negative $m = \dfrac{-3}{2}$

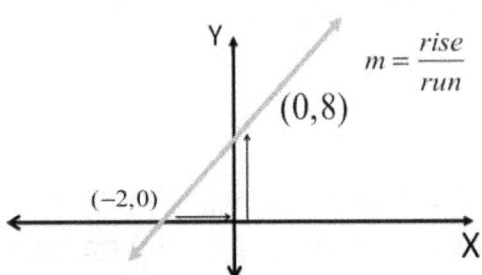

$m = \dfrac{rise}{run} = \dfrac{8}{2} = 4$, since the slope is positive $m = \dfrac{8}{2} = 4$

Let's Practice!

Identify the slope by the Rise/Run method

1)

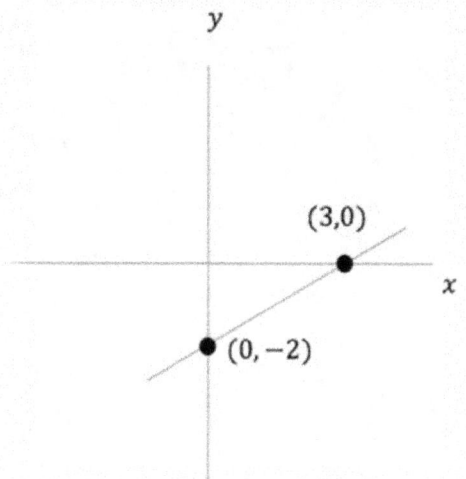

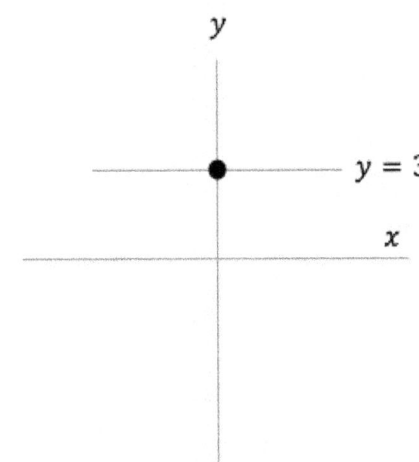

2)

3)

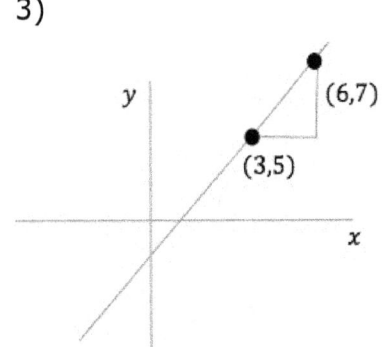

4)

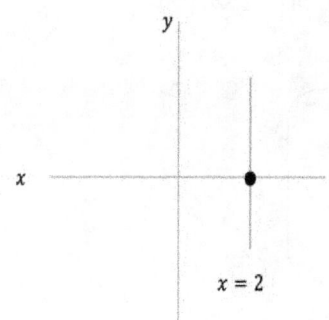

$x = 2$

5.

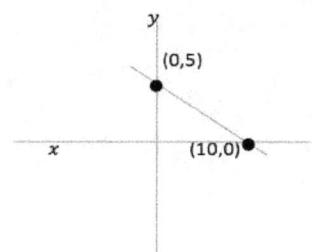

Let's check your answers

1) $m = \frac{2}{3}$

2) $m = 0$

3) $m = undefined$

5) $m = \frac{2}{3}$

4) $m = \frac{-5}{10} = \frac{-1}{2}$

The slope of a line can also be found by using the slope formula when you are given two points (x_1, y_1) and (x_2, y_2).

SLOPE FORMULA
$$m = \frac{y_2 - y_1}{x_2 - x_1}$$

Find the slope of the line with the following points:

$$(3, -2)\ (5, -1) \implies m = \frac{-1-(-2)}{5-3} = \frac{-1+2}{2} = \frac{1}{2}$$

Let's Practice!

Find the slope of the following points

1. $(2, 5)\quad (-1, 3)$
2. $(6, -1)\quad (3, -1)$
3. $(16, 2)\quad (8, -7)$
4. $(5, 0)\quad (5, 3)$
5. $(10, 13)\quad (-10, -13)$

Let's check your answers

1. $(2, 5)\ (-1, 3) \implies m = \frac{3-5}{-1-2} = \frac{-2}{-3} = \frac{2}{3}$

2. $(6, -1)\ (3, -1) \implies m = \frac{-1-(-1)}{3-6} = \frac{-1+1}{-3} = \frac{0}{-3} = 0$

3. $(16, 2)\ (8, -7) \implies m = \frac{-7-2}{8-16} = \frac{-9}{-8} = \frac{9}{8}$

4. $(5, 0)\ (5, 3) \implies m = \frac{3-0}{5-5} = \frac{3}{0} = \infty$ (Undefined)

5. $(10, 13)\ (-10, -13) \implies m = \frac{-13-13}{-10-10} = \frac{-26}{-20} = \frac{13}{10}$

Intercepts

Lines and curves will intercept at the y-axis or x-axis.
For example:

If $y = -\frac{3}{2}x + 3,$ find the intercepts:

In order to **find the x- axis intercept, make $y = 0$.**
Solve for x.

$$y = -\frac{3}{2}x + 3$$

$$0 = -\frac{3}{2}x + 3$$

$$-3 = -\frac{3}{2}x$$

$$-6 = -3x$$

$$x = \frac{-6}{-3} = 2$$

So, the x-axis intercept is (2,0)

In order to find **the y-axis intercept, make $x = 0$.**
Solve for y.

$$y = -\frac{3}{2}x + 3$$

$$y = -\frac{3}{2}(0) + 3$$

$$y = 0 + 3$$

$$y = 3$$

The y - axis intercept is (0,3)

As seen in the graph, the intercepts are the values when one coordinate is zero the other one is the intercept.

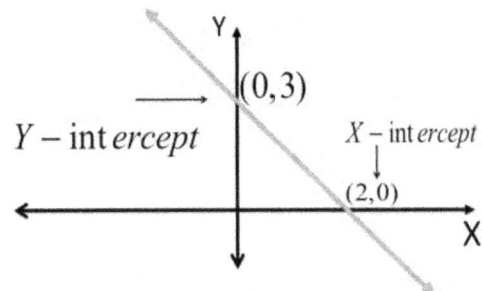

Let's Practice!

Find the X-intercepts   (Hint: make y=0)
1. $2x + 6y = 12$
2. $3x - 5y + 2 = 0$
3. $x - 3 = y$

Find the y-intercepts
1. $x - y = 5$
2. $10x - 5y = -20$
3. $y = 4x + 3$
4. $y = 3x - 2$

X-intercepts

1. $2x + 6y = 12$

   $2x + 6(0) = 12$
   $2x = 12$
   $x = 6,$ $(6,0)$

2. $3x - 5y + 2 = 0$
   $3x - 5(0) + 2 = 0$

$3x = -2$

$x = \frac{-2}{3}$ $\left(\frac{-2}{3}, 0\right)$

3. $x - 3 = y$
   $x - 3 = 0$
   $x = 3$ (3,0)

Y-intercepts

1. $x - y = 5$
   $0 - y = 5$
   $y = -5$
   $(0, -5)$

2. $10x - 5y = -20$
   $10(0) - 5y = -20$
   $y\frac{-20}{-5} = 4$
   $(0, 4)$

3. $y = 4x + 3$
   $y = 4(0) + 3$
   $y = 3$
   $(0, 3)$

4. $y = 3x - 2$
   $y = 3(0) - 2$
   $y = -2$
   $(0, -2)$

Building the Equation of the Line

The point-slope formula lets you build the equation of the line, you just need the slope and one point:

> **POINT – SLOPE FORMULA**
>
> $y - y_1 = m(x - x_1)$

For example:

If you have a line with $m = 3$ (slope = 3), that passes through point (1,2), then we can build the equation of the line by substituting the point and the slope into the formula.

$$y - y_1 = m(x - x_1)$$

$$y - 2 = 3(x - 1)$$

$$y - 2 = 3x - 3$$

$$y = 3x - 3 + 2$$

$$y = 3x - 1$$

The final answer is $y = 3x - 1$

You can also use the $y = mx + b$ formula to build the equation of the line.

These are the most common forms of the line.

SLOPE INTERCEPT FORM
$y = mx + b$
b = y-axis intercept
m = **slope**

STANDARD FORM
$ax + by = c$
POINT-SLOPE FORM
$y - y_1 = m(x - y_2)$

Let's do an example using the slope intercept form:
- **Build the equation of the line that passes through point (1, 2) with $m=3$.**

Since you have (x,y) and m, let's substitute it into the equation $y = mx + b$ to find b. Remember, **b is also the y-intercept**.

$$y = mx + b$$
$$2 = (3)1 + b$$
$$2 = 3 + b$$
$$b = 2 - 3$$
$$b = -1$$

Finally, we can substitute it into $y = mx + b$, since $b = -1$, then
$$y = 3x - 1$$

Very Important:
You can use both methods. They both will give you the same answer. **Remember you only need the slope and one point**.

Let's do another example with two coordinates:

Build the equation of the line that passes through (3,-1) and (2, 5).

STEP ONE: In this example, you need to find the slope first.

$$m = \frac{Y2-Y1}{x2-x1} = \frac{5-(-1)}{2-3} = \frac{5+1}{-1} = \frac{6}{-1} = -6$$

Since the slope is m=-6, then we just need one coordinate to build the equation of the line. You can choose (3,-1) or (2, 5). In this case let's choose (3,-1).

STEP TWO: Solve for the value of b

$$y = mx + b$$

-1 = (-6)(3) + b
-1 = -18 + b
b = -1 + 18
b = 17

Finally, just remember your slope is m=-6 and b=17, then the slope intercept form is:

$$y = -6x + 17$$

Let's Practice!

Build the equation of the line
1. $(3,5)\ (2,6)$
2. $m = 5,\ (8,0)$
3. $(2,-3)\ (0,5)$
4. $m = \frac{-2}{3},\ (13,5)$
5. $m = undefined,\ (2,5)$
6. $m = 0,\ (5,6)$

Let's check your answers

1. $m = \frac{6-5}{2-3} = \frac{1}{-1} = -1$

   Pick any point, I will pick $(2,6)$ and use $y - y_1 = m(x - x_1)$
   $y - 6 = -1(x - 2)$
   $y - 6 = -x + 2$
   $y = -x + 2 + 6$
   $y = -x + 8$

2. $m = 5,\ (8,0)$
   $y - 0 = 5(x - 8)$
   $y = 5x - 40$

3. $(2,-3)\ (0,5)$

   $m = \frac{5-(-3)}{0-2} = \frac{5+3}{-2} = \frac{8}{-2} = -4$

   $y + 3 = -4(x - 2)$
   $y + 3 = -4x + 8$
   $y = -4x + 8 - 3$

$$y = -4x + 5$$

4. $m = \frac{-2}{3}, \quad (13,5)$

$$y - 5 = \frac{-2}{3}(x - 13)$$

$$y - 5 = \frac{-2x}{3} + \frac{26}{3}$$

$$y = \frac{-2x}{3} + \frac{26}{3} + 5$$

$$y = \frac{-2x}{3} + \frac{41}{3}$$

5. $m = undefined, \quad (2,5)$

   Since the slope is undefined, then $x = 2$

6. $m = 0, \quad (5,6)$
   $$y - 6 = 0(x - 5)$$
   $$y - 6 = 0$$
   $$y = 6$$

Graphing a Line

The easiest way to graph a line is to put in into slope-intercept form; $(y = mx + b)$ where m is the slope and b is the y-intercept.

Let's see some examples:
Let's say you have the line y = 3x − 4.

STEP ONE:

$$y = 3x - 4$$

Identify the components of the equation:

y = the value of y

$3 = m$ (the slope and the value next to the x)

$-4 = b$ (the y-intercept)

Then we have….

$$y = 3x - 4$$

Following the slope-intercept form we have:

$$y = 3x - 4 \implies y = mx + b$$

$m = 3$ (slope)

$b = -4$ (intercept with the y −axis)

STEP TWO:

$$y = 3x - 4$$

Find the y −intercept:

$$(0, -4)$$

This is because $b = -4$, that means when $x = 0, y = -4$

STEP THREE:

$$y = 3x - 4$$

Find the slope:

$m = 3$ or you can rewrite it as a fraction, like this: $m = \frac{3}{1}$

STEP FOUR: Graph the line

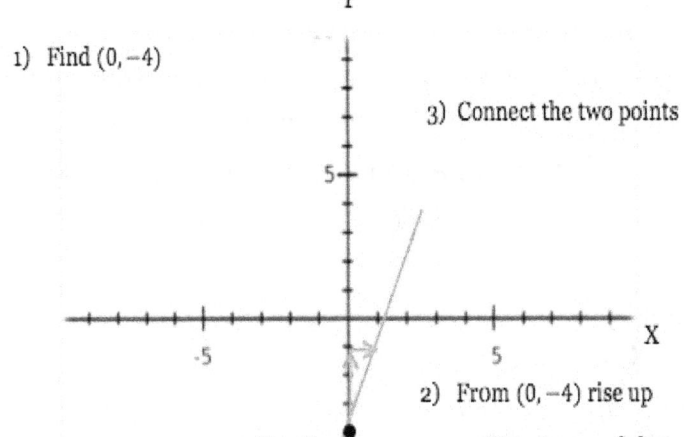

How to graph?

$y = 3x - 4$

1) Find $(0, -4)$

2) From $(0, -4)$ rise up (3) spaces, and then run (1) space to the right

3) Connect the two points

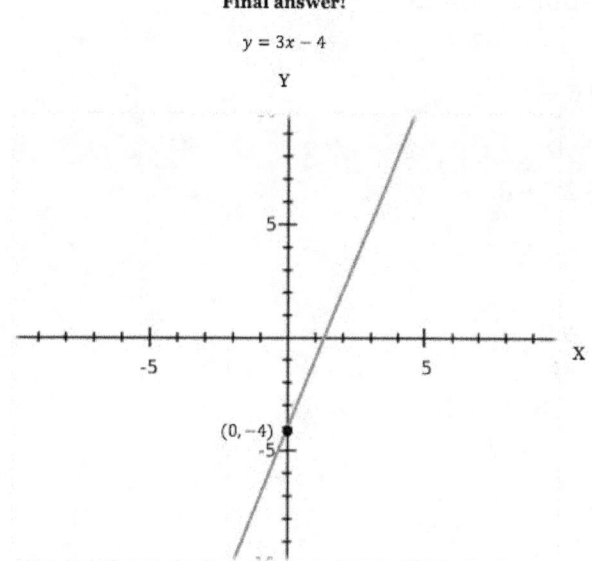

Let's Practice!

Graph the following lines (Hint: It is easier if you make your lines in the $y = mx + b$ format (that is to solve for y).

1. $2x + 4y = 8$
2. $y = \frac{-3}{2}x + 5$
3. $3x = -2y + 6$
4. $x = 3 - y$
5. $2x = 6y + 10$
6. $5x = 2y$
7. $y = 5$
8. $x = -2$

Let's check your answers:

1. $2x + 4y = 8$

$$4y = -2x + 8$$

$$\frac{4y}{4} = \frac{-2x}{4} + \frac{8}{4}$$

$$y = \frac{-x}{2} + 2$$

$m = \frac{-1}{2}$ and $b = 2$

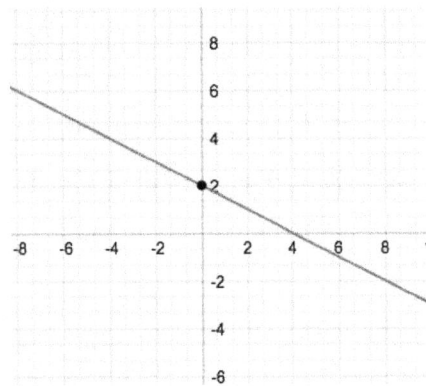

2. $y = -\frac{-3x}{2} + 5$

        $m = \frac{-3}{2}$ $b = 5$

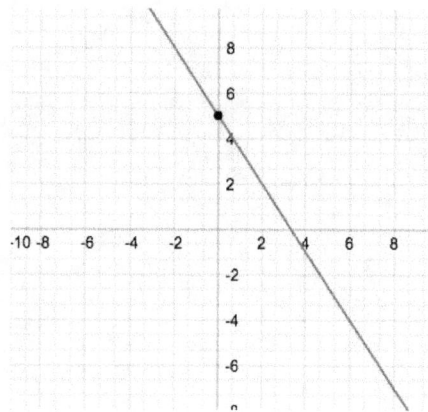

3. $3x = -2y + 6$

   $2y = -3x + 6$

   $y = -\frac{3x}{2} + 3$

   $m = \frac{-3}{2} \qquad b = 3$

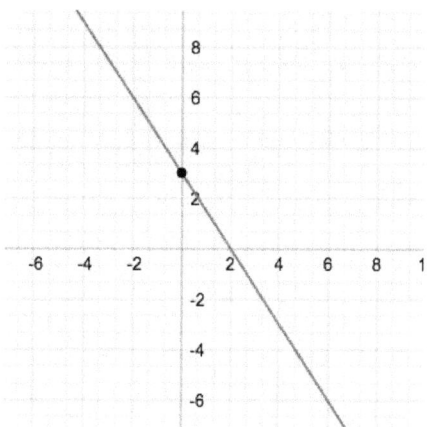

4. $x = 3 - y$

   $x - 3 = -y$

   $y = -x + 3$

   $m = \frac{-1}{1} \qquad b = 3$

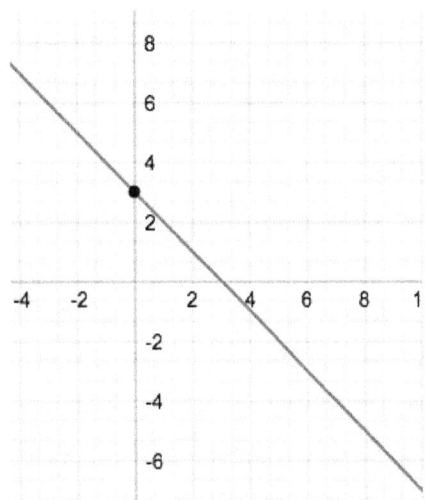

5. $2x = 6y + 10$

      $-6y = -2x + 10$

      $y = (-\frac{2}{-6})x + (10/-6)$

      $y = \frac{x}{3} - 5/3$

      $m = \frac{1}{3}$ $b = -\frac{5}{3}$

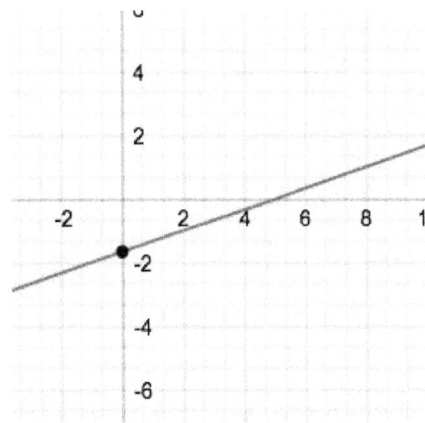

6. $5x = 2y$

   $-2y = -5x$

   $y = \left(-\dfrac{5}{-2}\right)x$

   $y = \dfrac{5x}{2}$

   $m = \dfrac{5}{2} \quad b = 0$

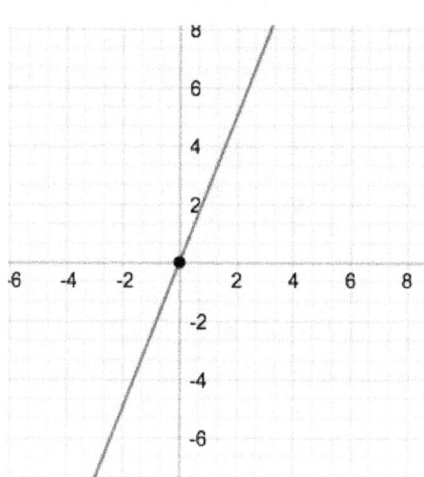

7. $y = 5$

   This is a special case where the slope (m) is zero (a horizontal line) and the y-intercept (b) is 5.

   $m = 0 \quad b = 5$

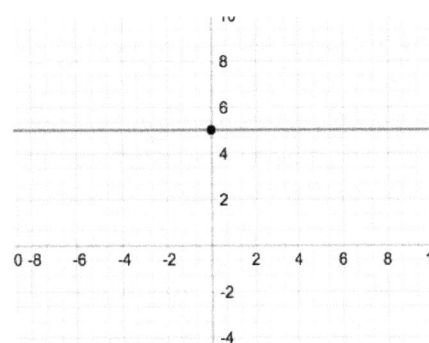

8. $x = -2$

This is a special case where the slope is undefined (a vertical line) and the y-intercept is zero

$m = Undefined\ (vertical\ line) \qquad b = 0$

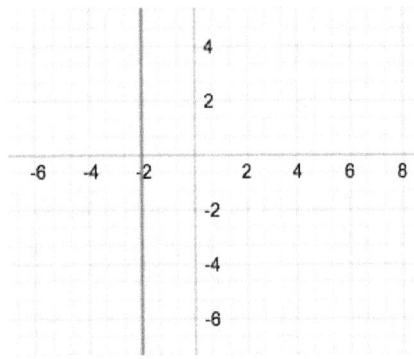

Parallel Lines

Two parallel lines have the same slope.

For example: $y = 3x - 2$ and $y = 3x + 5$ are parallel lines ($m = 3$).

They will never intersect each other.

$y = 3x + 5$

$y = 3x - 2$

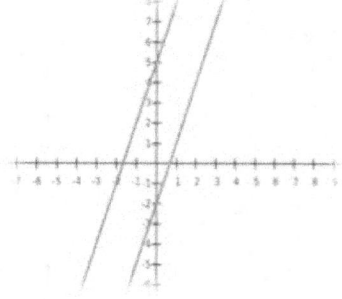

Let's say they ask you to find **the slope parallel to the line**

$2x - 5y = 10$.

STEP ONE: Set up the equation in slope-intercept form ($y = mx + b$).

$$y = \frac{2}{5}x - 2$$

STEP TWO: Then look at the equation and find the value of m.

m (slope) $= \frac{2}{5}$

This is the slope of the line parallel to the line $2x - 5y = 10$

Perpendicular Lines

Two lines are perpendicular when they intersect each other and make 90° angle when they intercept. The slopes have opposite signs and they are reciprocal to each other.

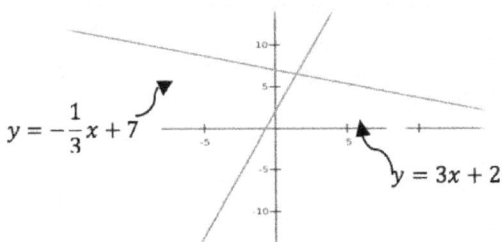

For example: $y = 3x + 2$ has a slope $m = 3$, the perpendicular slope would be $m \perp = \frac{-1}{3}$.

Another example:
Find the slope perpendicular to the line $3x - 10 = -5y$,

STEP ONE: Set up the equation in slope-intercept form: $y = mx + b$

$$y = \frac{-3}{5}x + 2$$

The perpendicular slope is $m \perp = \frac{5}{3}$ (negative and inverse)

> **The notation $m_\parallel$ means parallel and $m_\perp$ means perpendicular**

Let's Practice!

Find the slopes parallel and perpendicular to the following lines.

1. $2x + 6y = 1$
2. $x = 2y - 5$
3. $x = 3$
4. $y = 2$
5. $x = -8y$

Let's check your answers

1. $2x + 6y = 1$
   $6y = -2x + 1$
   $y = \frac{-2}{6}x + \frac{1}{6}$

   $m = \frac{-2}{6} = -\frac{1}{3}$

   $m_\parallel = -\frac{1}{3}$ $\quad m_\perp = 3$

2. $x = 2y - 5$
   $2y = x + 5$
   $y = \frac{x}{2} + \frac{5}{2}$

   $m = \frac{1}{2}$

   $m_\parallel = \frac{1}{2}$ $\quad m_\perp = -2$

3. $x = 3 \Longrightarrow m = 0$

   $m_\parallel = 0 \quad m_\perp = undefined$

   Hint: When the slope is zero, the perpendicular slope is undefined

4. $y = 2$

   $m = undefined$

   $m_\parallel = undefined \quad m_\perp = 0$

5. $x = -8y$

   $y = -\frac{x}{8}$

   $m = -\frac{1}{8}$

   $m_\parallel = -\frac{1}{8} \quad m_\perp = 8$

Building the Equation of Parallel & Perpendicular Lines

Build the equation of a line parallel to $2x - 3y = 6$ that passes through the points (3,5).

STEP ONE: Find the slope of the line

$$-3y = -2x + 6$$

$$y = \frac{2}{3}x + 2$$

$$m = \frac{2}{3}$$

Therefore, the slope of the parallel is $m_\parallel = \frac{2}{3}$

STEP TWO: Use the slope-intercept form to find b (y-intercept) using the given point (3,5)

Substitute all the values $y = 5$, $x = 3$, $m = \frac{2}{3}$

$$y = mx + b$$

$$5 = \frac{2}{3}(3) + b$$

$$5 = 2 + b$$

$$b = 3$$

STEP THREE: Finally, plug your slope and y-intercept to the line

$$y = mx + b$$

$$y = \frac{2}{3}x + 3$$

STEP FOUR: You can graph them to check your answer.

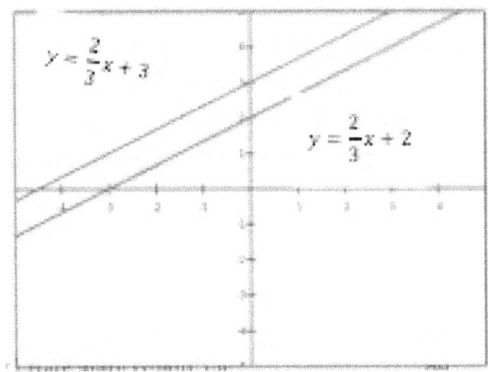

Let's do the same but with the perpendicular line

Build the equation of a line perpendicular to $2x - 3y = 6$ and that passes through the point (3, 5).

STEP ONE: Find the slope of $2x - 3y = 6$

$$-3y = -2x + 6$$

$$y = \frac{2}{3} - 2$$

$$m = \frac{2}{3}$$

Therefore, the slope of a line perpendicular will be (the inverse and opposite sign)

$$m_\perp = \frac{-3}{2}$$

STEP TWO: Use the equation of a line formula to find b (y-intercept)

$$m_\perp = \frac{-3}{2} \qquad (3,5)$$

$$y = mx + b$$

$$5 = \frac{-3}{2}(3) + b$$

$$5 = \frac{-9}{2} + b$$

$$b = \frac{9}{2} + 5$$

$$b = \frac{19}{2}$$

STEP THREE: Finally, plug your slope and $y - intercept$ to the line

$$y = mx + b$$
$$y = mx + b$$
$$y = \frac{-3}{2}x + \frac{19}{2}$$

Let's Practice!

1. Build the equation of the line parallel to $-3x = y - 6$ that passes through the point (2, 5).

2. Build the equation of the line perpendicular to $2x - 6y = 12$ that passes through the point (0,-3).

3. Build the equation of the line perpendicular to $-2y = x$ that passes through the point (-8,-10).

4. Build the equation of the line perpendicular to y=3 that passes through the point (4, 5).

5. Build the equation of the line parallel to y=3 that passes through the point (4, 5).

Let's check your answers:

1. $-3x = y - 6$

STEP ONE: Find the slope in the equation solving for y=mx+b.
$y = -3x + 6$

STEP TWO: Determine the slope is
$m=3$, then the parallel slope is also $m_\parallel = 3$,

STEP THREE: Now we can build the equation of the line with the point (-2, 5)
$y = mx + b$
5=(3)(-2)+b
5= -6 + b
$b = -11$
$y = 3x - 11$

2. $2x - 6y = 12$

STEP ONE: Find the slope in the equation 2x-6y=12 and solve for
$y = mx + b$
$Y = (\frac{x}{3}) - 2$

$m = (\frac{1}{3})$

STEP TWO: Determine the perpendicular slope (negative and inverse)
$m_\perp = -3$

STEP THREE: Now we can build the equation of the line that has slope $m=-3$ and passes through the point (0,-3)
$y = mx + b$

-3 = (-3)(0) + b
-3 = b
b = -3
y = -3x - 3

3. $-2y = x$

STEP ONE: Find the slope in the equation $-2y = x$ by solving for y
$y = -\frac{x}{2}$
$m = -\frac{1}{2}$

STEP TWO: Determine the perpendicular slope
$m_\perp = 2$

STEP THREE: Now we can build the equation of the line that has a slope $m = 2$ and passes through the point (-8,-10)

$y = mx + b$
-10 = (2)(-8) + b
-10 = -16 + b
b = 6
$y = 2x + 6$

4. $y = 3$

This is a special case. Remember the slope of $y = 3$ is $m = 0$. It's a horizontal line and the perpendicular slope is the vertical line.
Vertical lines have an undefined slope.
So, with a line that passes the point (4, 5) and is vertical, you will pick the x value to get the vertical line.
$x = 4$ is your answer!

5. $y = 3$

This is a special case. Remember the slope of $y = 3$ is $m = 0$.

It's a horizontal line. A parallel slope will also be a horizontal line. Then, with a line that passes the point (4, 5) and is horizontal, you will pick the y value to get the horizontal line.

$y = 5$ is your answer!

System of Equations: Solving by Graphing

Let's say you have two lines such as: $y = 2x + 5$ and $y = 3x + 4$. When these **two lines intersect**, they share a common point that is called the **Solution of the System of Equations**.
When you graph them, they share a **common point**.
In this case (1, 7) is the solution of the system of equations.

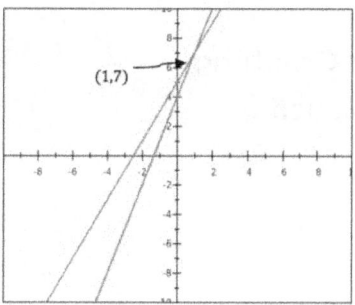

Let's Practice!

Find the solution of the system of equations

$$y = 2x + 5 \text{ and}$$
$$8x - 3y = -11$$

STEP ONE: First you need to arrange the equations as $y = mx + b$, $y = 2x + 5$ is done but, $8x - 3y = -11$ needs to be rearranged.

$$8x - 3y = -11$$
$$-3y = -8x - 11$$
$$\frac{-3y}{-3} = \frac{-8}{-3}x - \frac{11}{-3}$$

$$y = \frac{8}{3}x + \frac{11}{3}$$

STEP TWO: Now let's graph them and find their intersection

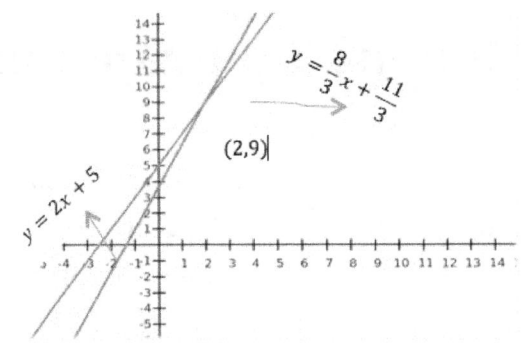

STEP THREE: The solution is (2, 9).

Let's Practice Some Graphing!

Graph and find the solution

1. $2x - y = 6,\quad y = 5x - 18$
2. $\frac{2}{3}x - y = 5,\quad 3x - 6y = 27$
3. $x = 3,\quad y = 2$
4. $y = 5x - 2,\quad 5y = 25x + 10$
5. $2x + 4y = 8,\quad x + 2 = 4$

Let's check your answers:

1. $\quad 2x - y = 6 \quad\bigg|\quad y = 5x - 18$
   $\quad -y = -2x + 6$
   $\quad y = 2x - 6 \quad\bigg|\quad y = 5x - 18$

   The answer is (4,2)

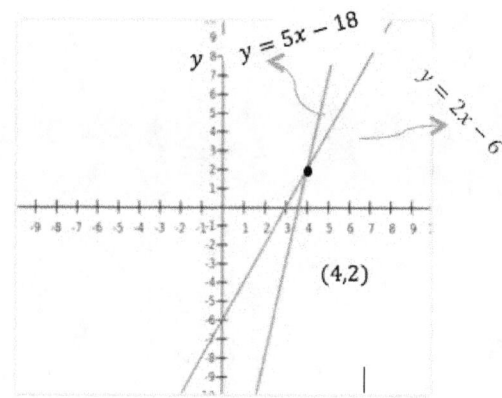

2. $\frac{2}{3}x - y = 5,$ $\quad\quad\quad 3x - 6y = 27$

$\frac{2}{3}x = 5 + y$ $\quad\quad\quad -6y = -3x + 27$

$y = \frac{2}{3}x - 5$ $\quad\quad\quad \frac{-6y}{-6} = \frac{-3x}{-6} + \frac{27}{-6}$

$\quad\quad\quad\quad\quad\quad\quad y = \frac{x}{2} - \frac{9}{2}$

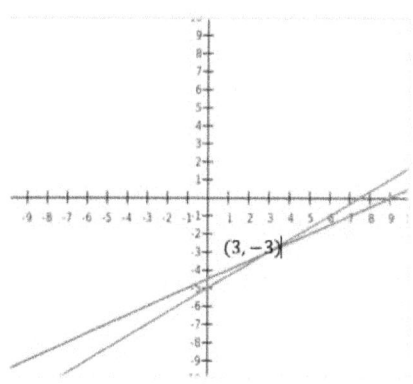

The answer is (3,-3)

3. $x = 3, \quad y = 2$

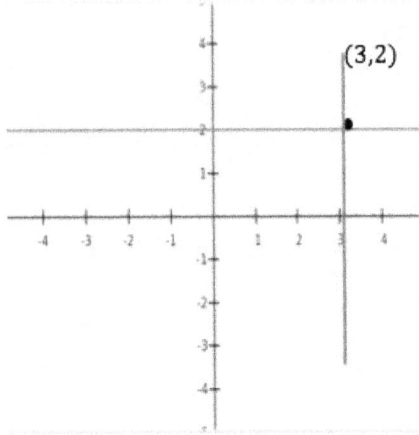

4) $5y = 25x + 10,$ $\qquad$ $y = 5x - 2$

$\dfrac{5y}{5} = \dfrac{25}{5}x + \dfrac{10}{5}$

$y = 5x + 2,$ $\qquad$ $y = 5x - 2$

There is no solution since both lines have the same slope (parallel) and they will never intersect each other.

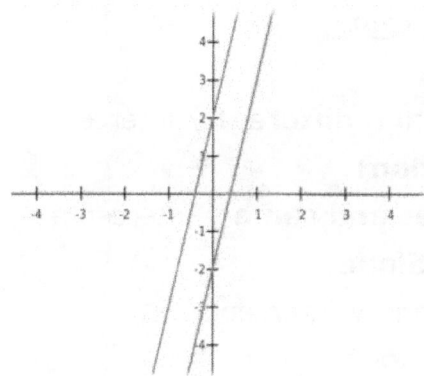

5. $\quad 2x + 4y = 8 \quad \bigg| \quad x + 2y = 4$

$ 4y = -2x + 8 \qquad 2y = -x + 4$

$\quad \dfrac{4y}{4} = \dfrac{-2x}{4} + \dfrac{8}{4} \quad \bigg| \quad \dfrac{2y}{2} = \dfrac{-x}{2} + \dfrac{8}{4}$

Both lines are identical; therefore, there are infinitive solutions.

$\quad y = \dfrac{-x}{2} + 2 \qquad\qquad y = \dfrac{-x}{2} + 2$

Important to know

1) If you have the **same slope and different** y-intercepts
(**No solution**)
2) If you have the **same slopes and same** y-intercepts
(**Infinitive Slope**)
3) **Different** slopes and y-intercepts
(**One solution**)

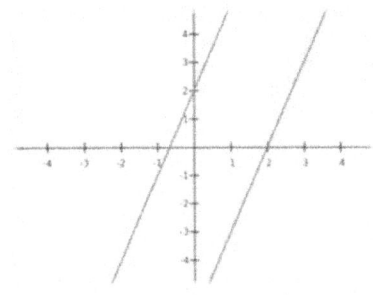 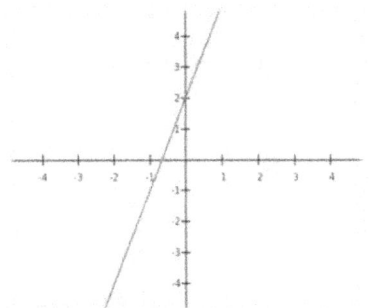 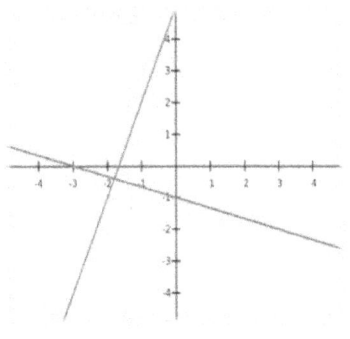

1. Example 1 2. Example 2 3. Example 3

$y = 3x + 2,$ $y = 3x + 2$ or $y = 3x + 5,$
$y = 3x - 6$ $y = 3x + 2$ $y = \frac{-x}{3} - 1$

No Solution Infinite Solutions One Solution

System of Equations Using Substitution

System of equations can also be solved algebraically.

For example:

$$2x - y = 6$$
and $y = x - 1$

STEP ONE: Find the easiest equation to substitute into the other equation. In this case $y = x - 1$ is solved, so I will plug it into

$$2x - y = 6.$$
$$2x - (x - 1) = 6$$
$$2x - x + 1 = 6$$
$$x = 6 - 1$$
$$x = 5$$

STEP TWO: Now I can find y, by substituting into $y = x - 1$
$$y = 5 - 1$$
$$y = 4$$
my answer is (5,4)

Tip: Coordinates are always written like (x, y). Just remember they are in alphabetical order.

Let's Practice!

Solve the system of equations by using the substitution method.

1. $y = 2x + 5$, $y - 1 = 3x$
2. $2x - y = 10$, $y = 5x$
3. $5x = 10$, $y = \frac{3}{2}x + 1$
4. $2x + y = 20$, $3x - y = 0$
5. $y = 5x + 5$, $2y - 10x = -20$

Let's check your answers

1. $y = 2x + 5$, $y - 1 = 3x$

$\Longrightarrow$ $y = 3x + 1$ Solving for Y and then make them equal

$$y = 2x + 5$$
$$3x + 1 = 2x + 5$$
$$3x - 2x = 5 - 1$$
$$x = 4$$

Then find y; you can pick any equation.

Let's do $y = 2x + 5$
$$y = 2(4) + 5$$
$$y = 8 + 5 = 13$$

The solution is (4,13)

2. $2x - y = 10$, $y = 5x$

STEP ONE: Plug $y = 5x$ into $2x - y = 10$

$2x - (5x) = 10$

$2x - 5x = 10$

$-3x = 10$

$x = \frac{-10}{3}$

STEP TWO: to find y, let's plug $x = \frac{-10}{3}$ into $y = 5x$

$$y = 5\left(\frac{-10}{3}\right) = \frac{-50}{3}$$

The solution is $\left(\frac{-10}{3}, \frac{-50}{3}\right)$

3. $5x = 10$, $y = \frac{3}{2}x + 1$

   STEP ONE: Solve for x the easiest equation $5x = 10$

   $x = \frac{10}{5} = 2$

   STEP TWO: Plug $x = 2$ into $y = \frac{3}{2}x + 1$

   $y = \frac{3}{2}(2) + 1$

   $y = 3 + 1 = 4$

   The solution is (2, 4).

4. $2x + y = 20$, $3x - y = 0$

   STEP ONE: Put both equations into slope-intercept form
   $2x + y = 20 \longrightarrow y = -2x + 20$
   $3x - y = 0 \longrightarrow y = 3x$

   STEP TWO: Now plug $y = 3x$ into $y = -2x + 20$
   $3x = -2x + 20$
   $5x = 20 \implies x = 4$
   $y = 3(4) = 12$
   The solution is (4,12)

5. $y = 5x + 5$

   $\frac{2y}{2} = \frac{10x}{2} + \frac{20}{2}$

Now let's plug $y = 5x + 5$ into $y = 5x + 10$

$5x + 5 = 5x + 10$

$5x - 5x = 10 - 5$

$10 = 5$ but $10 \neq 5$

NO SOLUTION. They are parallel.

System of Equations Using Elimination

Let's say you have a system of equations:
$$2x - y = 8$$
$$x + y = 4$$

You can solve this system by ELIMINATING one variable.

STEP ONE: Find which variable is easier to eliminate, for example, if you add the system <u>vertically,</u> can be eliminated. In this case (y) is the easier variable to eliminate.

$$2x - \cancel{y} = 8$$
$$x + \cancel{y} = 4$$

STEP TWO: Add the system vertically eliminating Y.

$$2x = 8$$
$$x = 4$$
$$\overline{3x = 12}$$

STEP THREE: Solve for x

$$x = \frac{12}{3} = 4$$

STEP FOUR: Plug back x into any of the equations. $x = 4$. Let's pick

$$x + y = 4$$
$$4 + y = 4$$
$$y = 4 - 4$$
$$y = 0$$

The solution is $(4, 0)$

Let's do another example
$2x - 3y = 16$
$x + y = 3$

In this example, we can see that the variables will not eliminate if we add them <u>vertically</u>. We need to rewrite one of the equations so they can eliminate.

Let's multiply the second equation by -2 (EVERY SINGLE TERM) so that we can ELIMINATE the "2x" in the other equation.
$-2(x + y = 3)$
$-2(x) - 2(y) = -2(3)$
$-2x - 2y = -6$

Now we can add them <u>vertically</u> to eliminate the value of x

$\cancel{2x} - 3y = 16$
$\underline{\cancel{-2x} - 2y = -6}$
$-5y = 10$

Solve for y

$y = \frac{10}{-5}$

$y = -2$

Plug back $y = -2$ in any of the original equations. Let's do it on $x + y = 3$!
$x - 2 = 3$
$x = 3 + 2$
$x = 5$

The solution is $(5, -2)$

Let's Practice!

Solve the following systems of equations using the elimination method:

1. $x + y = 10$
   $3x - y = 15$

2. $x - y = 12$
   $4x - y = 6$

3. $2x - 5y = 11$
   $3x - 4y = 6$

4. $3x - 5y = 10$
   $y = 5x - 2$

5. $x + y = 20$
   $2y = -2x + 40$

6. $2 + y = 8x$
   $-8x + y = 5$

7. $x - 3 = 0$
   $y = 2x - 5$

8. $5x - y = 10$
   $-2x + 3y = 5$

Let's check your answers

1. $x + y = 10$
   $3x - y = 15$ $\Rightarrow$ Add vertically and eliminate the

   $x + y = 10$
   $3x - y = 15$
   _____
   $4x = 25$
   $x = \frac{25}{4}$

Then, find y by substituting $x = \frac{25}{4}$ into any equation. You can pick $x = y = 10$.

$x + y = 10$

$\frac{25}{4} + y = 10$

$y = 10 - \frac{25}{4}$

$y = \frac{15}{4}$

The solution is $\left(\frac{25}{4}, \frac{15}{4}\right)$

2. $\begin{aligned} x - y &= 12 \\ 4x - y &= 6 \end{aligned}$ ⟹ Let's eliminate y by multiplying the top equation by -1

$(-1)x - (-1)y = (-1)12$

Now you can add vertically

$\begin{aligned} -x + y &= -12 \\ 4x - y &= 6 \\ \hline 3x &= 6 \end{aligned}$

Solve for y

$x = \frac{-6}{3}$

$x = -2$

Plug $x = -2$ into any equation, let's pick $x - y = 12$

$-2 - y = 12$

$-y = 12 + 2$

$y = -14$

The solution is $(-2, -14)$

3. $\begin{aligned} 2x - 5y &= 11 \\ 3x - 4y &= 6 \end{aligned}$ ⟹ In this case, multiply the top by -3 and the bottom by 2, that way you get $-6x$ and $6x$ respectively

$$-3(2x) - (-3)5y = -3(11)$$
$$2(3x) - 2(4y) = 2(6)$$

$$\begin{aligned}-6x + 15y &= -33\\ 6x - 8y &= 12\\ \hline 7y &= -21\end{aligned}$$

$$y = \frac{-21}{7} \implies y = -3$$

Plug $y = -3$ into $2x - 5(-3) = 11$

$2x + 15 = 11$

$2x = -4 \implies x = -2$

The solution is $(-2, -3)$

4. $3x - 5y = 10$
   $y = 5x - 2$ $\implies$ In this case, it might be easier to us substitution since y is already solved. Substitute $y = 5x - 2$ into $3x - 5y = 10$.

   $3x - 5(5x - 2) = 10$
   $3x - 25x + 10 = 10$
   $-22x = 0$

   $x = \frac{0}{-22} = 0$

   Solving for y, plug $x = 0$ into $y = 5x - 2$
   $y = 5(0) - 2 = -2$

   The solution is $(0, -2)$

5. $x + y = 20$
   $2y = -2x + 40$ $\implies$ **STEP ONE**: Set up the second equation by moving $2x$ to the left side

   STEP TWO: Multiply $x + y = 20$ by $-2x$ in order to eliminate the x-value

   $(-2)x + (-2)y = (-2)20$
   $2x + 2y = 40$

$$-2x - 2y = -40$$
$$2x + 2y = 40$$
$$0 = 0$$

⟹ **STEP THREE:** Let's add vertically

This system has **INFINITE SOLUTIONS** because 0=0.

6. $2 + y = 8x$
    $-8x + y = 5$

⟹ You need to arrange the system so that x and y values are on the same side

$$8x - y = 2$$
$$-8x + y = 5$$
$$0 = 7$$

Now you can add vertically.
Your answer is 0=7 but this is FALSE.
This system of equations has **NO SOLUTION**.

Lines are parallel, there is no solution.

7. $x - 3 = 0$
    $y = 2x - 5$

    $y = 2(3) - 5$
    $y = 6 - 5$
    $y = 1$

STEP ONE: Solve $x - 3 = 0$. What is x?
STEP TWO: Substitute the x-value

The solution is $(3, 1)$.

8. $5x - y = 10$
    $-2x + 3y = 5$

The best way to eliminate y is by multiplying $5x - y = 10$ by 3

$(3)5x - (3)y = (3)10$
$-2x + 3y = 5$

⟹

$$15x - 3y = 30$$
$$-2x + 3y = 5$$
$$13x = 35$$

$$x = \frac{35}{13}$$

206

Now find y, by substituting x into $5x - y = 10$

$5\left(\frac{35}{13}\right) - y = 10.$

| Multiply all terms by 13 |

$13 \times 5\left(\frac{35}{13}\right) - 13y = 13 \times 10$

$175 - 13y = 130$

$-13y = 130 - 175$

$-13y = -45$

$y = \frac{-45}{-13} = \frac{45}{13}$

The solution is $\left(\frac{35}{13}, \frac{45}{13}\right)$

Graphing Linear inequalities

Graphing linear equations is very easy.
Let's take a look at the following rules first.

> \> or ≥ you shade **ABOVE** the line.
> (less than or less than or equal to)
> <, or ≤ you shade **BELOW** the line.
> < or > $\Rightarrow$ The line is dotted.
> ≤ or ≥ $\Rightarrow$ The line is Solid.

Let's do an example:
Graph the following linear inequality $y \geq 2x + 5$
STEP ONE: Graph the equation as a line, where $m = 2$ and $b = 5$
STEP TWO: Find out where to shade, either above or below the line.
$$y \geq 2x + 5$$

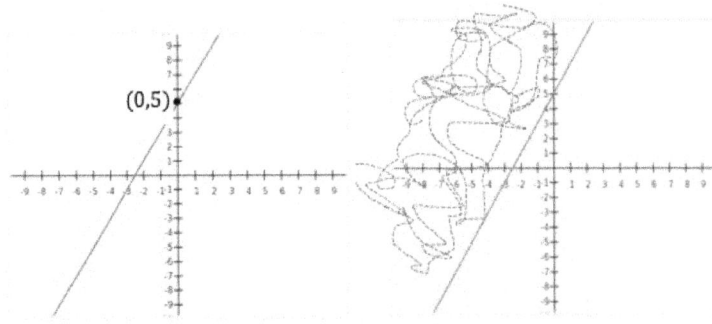

Let's do some examples:

1. $y < 3x + 2$ Dotted line

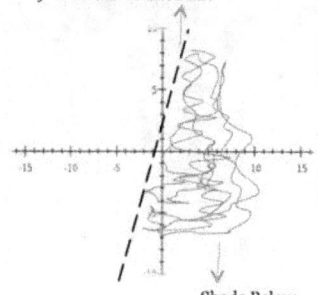

Shade Below

2. $2x + y \geq 6$
   $y \geq -2x + 6$
   Solid line   Shade above

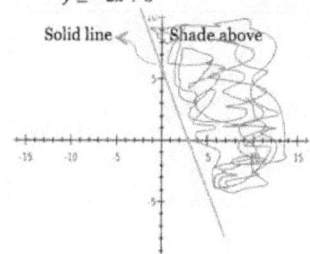

3. $x \geq -3$

Solid Line   Shade Right

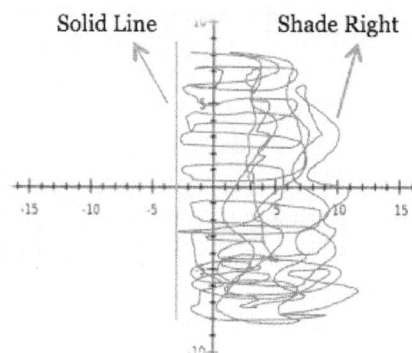

4. $y < 2$

Dotted line

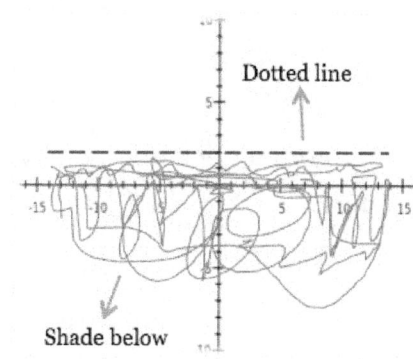

Shade below

5. $-y + 2x < 6$

$-y < -2x + 6$

$y > 2x - 6$

Since the negative, the sign flips

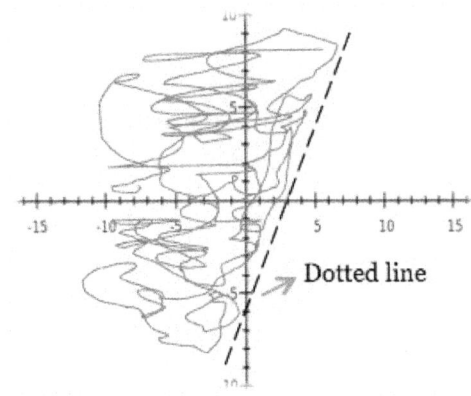

Dotted line

Let's Practice!

1. $8x + y < 10$

2. $-3y \leq 6$

3. $-x + y \geq 5$

4. $x + 5 < 0$

Let's check your answers

1. $8x + y < 10$

   $y < -8x + 10$

2. $-3y \leq 6$

   $y \geq \frac{6}{-3}$

   $y \geq -2$

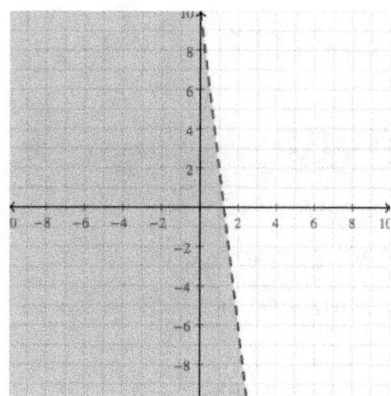

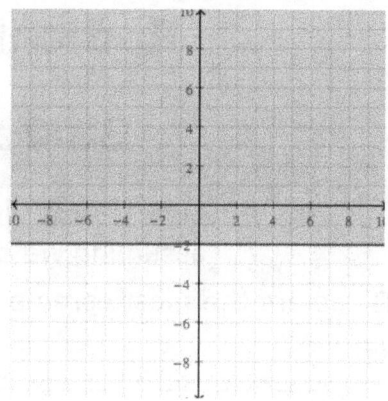

3. $-x + y \geq 5$
    $y \geq x + 5$

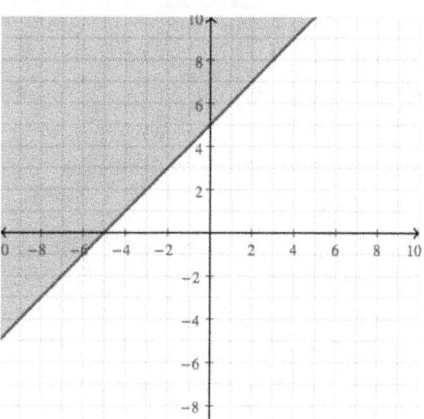

4. $x + 5 < 0$
    $x < -5$

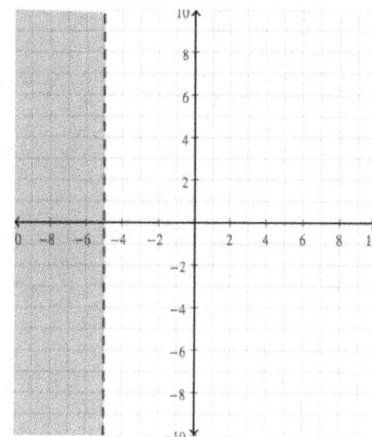

Radical Expressions:
Basic Operations

Let's say you have $\sqrt{4}$, the answer is 2
(assuming we only want positive answers).
These are the most common radical expressions or perfect squares:

$$\sqrt{1} = 1 \qquad \sqrt{4} = 2 \qquad \sqrt{9} = 3 \qquad \sqrt{16} = 4$$
$$\sqrt{25} = 5 \qquad \sqrt{36} = 6$$
$$\sqrt{49} = 7 \qquad \sqrt{64} = 8 \qquad \sqrt{81} = 9 \qquad \sqrt{100} = 10$$

Let's simplify $\sqrt{72}$, you can re-write $\sqrt{72}$ with <u>perfect squares</u>.

$$\sqrt{72} = \sqrt{9} \times \sqrt{4} \times \sqrt{2} = 3 \times 2 \times \sqrt{2} = 6\sqrt{2}$$

Let's Practice!

1. $\sqrt{192} - \sqrt{108}$
2. $\sqrt{5}(\sqrt{7} + \sqrt{5} - \sqrt{2})$
3. $(3 + \sqrt{2})(3 - \sqrt{2})$
4. $\sqrt{6}(\sqrt{3} - \sqrt{2}) + 7\sqrt{3}$
5. $\sqrt{180} - 4\sqrt{80}$
6. $\sqrt{\frac{120}{2}} + 6\sqrt{135}$

Let's check your answers

1. $\sqrt{192} - \sqrt{108} = \sqrt{64} \times \sqrt{3} - \sqrt{36} \times \sqrt{3} = 8\sqrt{3} - 6\sqrt{3} = 2\sqrt{3}$
2. $\sqrt{5}(\sqrt{5} + \sqrt{7} - \sqrt{2}) = \sqrt{25} + \sqrt{35} - \sqrt{10} = 5 + \sqrt{35} - \sqrt{10}$
3. $(3 + \sqrt{2})(3 - \sqrt{2}) = 9 - 3\sqrt{2} + 3\sqrt{2} - \sqrt{4} = 9 - 2 = 7$
4. $\sqrt{6}(\sqrt{3} - \sqrt{2}) + 7\sqrt{3} = \sqrt{18} - \sqrt{12} + 7\sqrt{3} = \sqrt{9} \times \sqrt{2} - \sqrt{4} \times \sqrt{3} + 7\sqrt{3}$

    $= 3\sqrt{2} - 2\sqrt{3} + 7\sqrt{3} = 3\sqrt{2} + 5\sqrt{3}$

5. $\sqrt{180} - 4\sqrt{80} = \sqrt{36} \times \sqrt{5} - 4\sqrt{16} \times \sqrt{5}$

    $= 6\sqrt{5} - 4 \times 4\sqrt{5} = 6\sqrt{5} - 16\sqrt{5} = -10\sqrt{5}$

6. $\sqrt{\frac{120}{2}} + 6\sqrt{135} = \sqrt{60} + 6\sqrt{135} = \sqrt{15} \times \sqrt{4} + 6\sqrt{9}\sqrt{15}$

    $= 2\sqrt{15} + 6 \times 3 \times \sqrt{15} = 2\sqrt{15} + 18\sqrt{15} = 20\sqrt{15}$

Rationalizing Radicals

Let's say you have an expression like this one:

$$\frac{3}{\sqrt{2}}$$

You need to **rationalize the denominator**.
This means to **eliminate the radical in the denominator**.
You can do this by multiplying the numerator (top) and denominator (bottom) by the same radical in the denominator.

$$\frac{3(\sqrt{2})}{\sqrt{2}(\sqrt{2})} = \frac{3\sqrt{2}}{\sqrt{4}} = \frac{3\sqrt{2}}{2}$$

However, if you have an expression like this one: $\frac{3}{\sqrt{2}+3}$, then you can multiply the denominator by using the conjugate ($\sqrt{2} - 3$). The **conjugate** is the same expression with a different sign for the constant.

$$\frac{3}{\sqrt{2}+3} \frac{(\sqrt{2}-3)}{(\sqrt{2}-3)} = \frac{3\sqrt{2}-9}{\sqrt{2}\sqrt{2}-3\sqrt{2}+3\sqrt{2}-3(3)} = \frac{3\sqrt{2}-9}{\sqrt{4}-9}$$

$$\frac{3\sqrt{2}-9}{2-9} = \frac{3\sqrt{2}-9}{-7} \quad or \quad \frac{-3\sqrt{2}-9}{7}$$

Let's Practice!
Rationalize the following expressions

1. $\dfrac{8}{\sqrt{3}}$ 2. $\dfrac{3}{\sqrt{5}-2}$ 3. $-\dfrac{5}{\sqrt{2}}$

4. $\dfrac{5}{5+\sqrt{3}}$ 5. $\dfrac{3+\sqrt{5}}{\sqrt{6}}$

Let's check your answers

1. $\dfrac{8}{\sqrt{3}} \times \dfrac{\sqrt{3}}{\sqrt{3}} = \dfrac{8\sqrt{3}}{\sqrt{9}} = \dfrac{8\sqrt{3}}{3}$

2. $\dfrac{3}{\sqrt{5}-2} \times \dfrac{\sqrt{5}+2}{\sqrt{5}+2}$ you need to multiply by the conjugate ($\sqrt{5}+2$)

$\dfrac{3\sqrt{5}+6}{\sqrt{5}\times\sqrt{5}+2\sqrt{5}-2\sqrt{5}-4} = \dfrac{3\sqrt{5}+6}{5-4} = \dfrac{3\sqrt{5}+6}{1} = 3\sqrt{5}+6$

3. $-\dfrac{5}{\sqrt{2}} \times \dfrac{\sqrt{2}}{\sqrt{2}} = -\dfrac{5\sqrt{2}}{\sqrt{4}} = -\dfrac{5\sqrt{2}}{2}$

4. $\dfrac{5}{5+\sqrt{3}} \times \dfrac{5-\sqrt{3}}{5-\sqrt{3}} = \dfrac{25-5\sqrt{3}}{25+5\sqrt{3}-5\sqrt{3}-\sqrt{9}} = \dfrac{25-5\sqrt{3}}{25-3} = \dfrac{25-5\sqrt{3}}{22}$

5. $\dfrac{3+\sqrt{5}}{\sqrt{6}} \times \dfrac{\sqrt{6}}{\sqrt{6}} = \dfrac{3\sqrt{6}+\sqrt{30}}{\sqrt{36}} = \dfrac{3\sqrt{6}+\sqrt{30}}{6}$

Radical Expressions with Variables

Let's say you have a radical expression with variables ($\sqrt{a^6}$).

We are going to assume all variables are positive.

Remember, the square root ($\sqrt{x}$) has an invisible 2 in the index.

Like: $\sqrt[2]{a^6}$

You can rewrite this expression as $\sqrt[2]{a^6}$ as $a^{\frac{6}{2}}$, that is a^3

Let's do some examples:

1. $\sqrt{a^{10}} = a^{\frac{10}{2}} = a^5$
2. $\sqrt{x^{12}} = x^{\frac{12}{2}} = x^6$
2. $\sqrt{x^2} = x^{\frac{2}{2}} = x$
3. $\sqrt{x^{20}} = x^{\frac{20}{2}} = x^{10}$

If you have an expression with an odd exponent $\sqrt{x^7}$, then you rewrite it as $\sqrt{x^6 \cdot x}$ and then simplify $x^{\frac{6}{2}} \cdot x^{\frac{1}{2}} = x^3 \cdot x^{\frac{1}{2}}$.

Since you do not want fractions, then keep $x^{\frac{1}{2}}$ as a radical like this:

$$x^3 \cdot x^{\frac{1}{2}} = x^3 \sqrt{x}$$

Let's Practice!

1. $\sqrt{x^{12}y^{17}z^{21}}$ 2. $\sqrt{12x^{16}y^{11}}$

3. $\sqrt{18x^3} + x\sqrt{72x}$ 4. $\sqrt{75a^5} - \sqrt{27a^7}$

Let's check your answers

1. $\sqrt{x^{12}y^{17}z^{21}} = \sqrt{x^{12}y^{16}yz^{20}z} = x^{\frac{12}{2}}y^{\frac{16}{2}}z^{\frac{20}{2}} = \mathbf{x^6 y^8 z^{10} \sqrt{yz}}$

2. $\sqrt{12x^{16}y^{11}} = \sqrt{4*3x^{16}y^{10}y} = 2x^{\frac{16}{2}}y^{\frac{10}{2}} = \mathbf{2x^8 z^{10} \sqrt{3y}}$

3. $\sqrt{18x^3} + x\sqrt{72x} = \sqrt{9*2x^2 x} + x\sqrt{36*2x} = 3x\sqrt{2x} + 6x\sqrt{2x} = \mathbf{9x\sqrt{2x}}$

4. $\sqrt{75a^5} - \sqrt{27a^7} = \sqrt{75a^5} - \sqrt{27a^7} = \sqrt{25*3a^4 a} - \sqrt{9*3a^6 a} = 5a^{4/2}\sqrt{3a} - 3a^{6/2}\sqrt{3a} = \mathbf{5a^2\sqrt{3a} - 3a^3\sqrt{3a}}$

Radical Equations

Let's talk first about <u>extraneous solutions</u>.
These solutions make sense algebraically but they do not make the original equation true.

<u>Let's do an example:</u>
Solve the following equation and check for extraneous solutions

$$\sqrt{X-1} = X - 7$$

<u>Step:</u>   Square both sides to eliminate the square root.

$$\left(\sqrt{x-1}\right)^2 = (x-7)^2$$

$$X - 1 = (X-7)^2 \quad \Rightarrow \quad X - 1 = X^2 - 14X + 49$$

<u>Step 2:</u>   Move all the terms to make the equation equal to zero.

$$X^2 - 14X - X + 49 + 1 = 0$$
$$X^2 - 15X + 50 = 0$$

<u>Step 3:</u>   Factor or use quadratic formula.

$$X^2 - 15X + 50 = 0 \qquad (X-10)(X-5) = 0$$
$$X = 10 \quad \text{and} \quad X = 5$$

<u>Step 4:</u>   Check for extraneous solutions by plugging them back into the original equation.

$$\sqrt{X-1} = X - 7 \qquad\qquad \sqrt{X-1} = X - 7$$
$$\sqrt{10-1} = 10 - 7 \qquad\qquad \sqrt{5-1} = 5 - 7$$
$$\sqrt{9} = 3 \qquad\qquad\qquad \sqrt{4} = -2$$

$$3 = 3 \qquad\qquad 2 \neq -2$$

Yes, $X = 10$ is **a solution** and
$X = 5$ is an **extraneous solution**.

Let's practice:

Solve the following radical equations.

(A). $\sqrt{X + 3} = X - 3$ (B). $\sqrt{2X + 6} - 6 = 0$

(C). $\sqrt{3X + 1} = 2X - 6$ (D). $\sqrt{-4X - 7} = \sqrt{X + 13}$

Let's check your answers:

(A). $\sqrt{X + 3} = X - 3$ $(\sqrt{X + 3})^2 = (X - 3)^2$

$$X + 3 = (X - 3)^2$$
$$X + 3 = X^2 - 6X + 9$$
$$X^2 - 7X + 6 = 0$$
$$(X - 6)(X - 1) = 0$$
$$X = 6 \quad \text{and} \quad X = 1$$

Checking for extraneous Solutions:

$\sqrt{6 + 3} = 6 - 3$ $3 = 3$ $X = 6$ is a solution.

$\sqrt{1 + 3} = 1 - 3$ $\sqrt{4} = -2$

                   $2 \neq -2$

$X = 1$ is an extraneous solution.

(B). $\sqrt{2X + 6} - 6 = 0$ $\sqrt{2X + 6} =$

                     $(\sqrt{2X + 6})^2 = (6)^2$

$2X + 6 = 36$ $2X = 36 - 6$

$$2X = 30$$
$$X = 15$$

Checking for extraneous Solutions:

$\sqrt{2(15) + 6} - 6 = 0$ $\sqrt{36} - 6 = 0$ $6 - 6 = 0$

$X = 15$ is a solution.

(C). $\sqrt{3X+1} = 2X - 6$ $\qquad$ $(\sqrt{3X+1})^2 = (2X-6)^2$

$$3X + 1 = 4X^2 - 24X + 36$$

$4X^2 - 27X + 35 = 0$ $\qquad$ $(4X - 7)(X - 5) = 0$

$$X = 5 \quad \text{and} \quad X = \frac{7}{4}$$

Checking for extraneous Solutions:

$\sqrt{3(5) + 1} = 2(5) - 6 \qquad \sqrt{16} = 4$

$\qquad\qquad\qquad 4 = 4 \qquad X = 5$ is a solution.

$\sqrt{3\left(\frac{7}{4}\right) + 1} = 2\left(\frac{7}{4}\right) - 6 \qquad \sqrt{\frac{25}{4}} = -\frac{5}{2} \qquad \frac{5}{2} \neq \frac{-5}{2}$

$X = \frac{7}{4}$ is an extraneous solution.

(D). $\sqrt{-4X - 7} = \sqrt{X + 13}$

$$(\sqrt{-4X - 7})^2 = (\sqrt{X + 13})^2$$

$$-4X - 7 = X + 13$$

$$-5X = 20$$

$$X = -4$$

Checking for extraneous Solutions:

$\sqrt{-4(-4) - 7} = \sqrt{-4 + 13} \qquad\qquad \sqrt{16 - 7} = \sqrt{-4 + 13}$

$\qquad\qquad\qquad\qquad \sqrt{11} = \sqrt{11}$

$\qquad\qquad X = -4$ is a solution.

Quadratic Equations

Let's say you have $x^2 + 11x + 30 = 0$.

This is a **quadratic equation.**

In order to solve it you need to factor first

$x^2 + 11x + 30 = 0 \implies (x + 6)(x + 5) = 0.$

Then separate each factor and make it equal to zero.

$x + 6 = 0, \quad x + 5 = 0$

and then solve:

$x = -6, x = -5$

Your answer (also called **zeros** or **roots of quadratic equation**) is

$\{-6, -5\}$

> YOU MUST ALWAYS HAVE THE QUADRATIC EQUATION EQUAL TO ZERO, THEN YOU CAN FACTOR

> STANDARD FORM FOR QUADRATIC EQUATION
> $ax^2 + bx + c = 0$

Let's do some examples

1. $2x^2 = -10x$

   $2x^2 + 10x = 0$

   $2x(x + 5) = 0$

$$2x = 0 \qquad x + 5 = 0$$
$$x = 0 \qquad x = -5$$
$$\{-5, 0\}$$

2. $3x = -14x^2 + 2$
$14x^2 + 3x - 2 = 0$

| $14 \times 2 = 28$ |
| $7 - 4 = 3$ |

$14x^2 + 7x - 4x - 2 = 0$
$7x(2x + 1) - 2(2x + 1) = 0$
$(7x - 2)(2x + 1) = 0$

$7x - 2 = 0 \qquad 2x + 1 = 0$
$7x = 2 \qquad 2x = -1$
$x = \frac{2}{7} \qquad x = \frac{-1}{2}$

$\{\frac{-1}{2}, \frac{2}{7}\}$

STEP ONE: Start by arranging the quadratic function equal to zero.
STEP TWO: Remember in order to factor a polynomial, you must find the values that when multiply = 28 and added/subtracted = 3
STEP THREE: Once you factor you can then set each factor equal to zero and solve for x.
NOTE: Always make the ax^2 term positive, this way it is easier to factor. In this case ax^2 = -14x^2.

Let's Practice!
1. $-4x - 32 = -x^2$
2. $x^2 - 25 = 0$
3. $2x^2 + 3x - 35 = 0$
4. $(3x - 1)^2 = 0$
5. $x(5 - x) = 0$

Let's check your answers

1. $-4x - 32 = -x^2$
$x^2 - 4x - 32 = 0$
$(x - 8)(x + 4) = 0$

2. $x^2 - 25 = 0$
$(x + 5)(x - 5) = 0$
$x = -5, x = 5$

$x = 8, x = -4$ $\quad\quad\quad\quad\quad\quad\quad\quad \{-5,5\}$

$\{-4,8\}$

3. $\quad 2x^2 + 3x - 35 = 0$ $\quad\quad\quad\quad$ 4. $\quad (3x-1)^2 = 0$

You can square root both sides

$\boxed{\begin{array}{c} 2 \times 35 = 70 \\ 10 - 7 = 3 \end{array}}$ $\quad\quad\quad\quad \sqrt{(3x-1)^2} = \sqrt{0}$

$2x^2 + 10x - 7x - 35 = 0$ $\quad\quad\quad 3x - 1 = 0$

$2x(x+5) - 7(x+5) = 0$ $\quad\quad\quad 3x = 1$

$(2x - 7)(x + 5 = 0)$ $\quad\quad\quad\quad\quad x = \frac{1}{3}$

$2x - 7 = 0 \quad\quad x + 5 = 0$ $\quad\quad\quad \left\{\frac{1}{3}\right\}$

$2x = 7$

$x = \frac{7}{2} \quad\quad\quad\quad x = -5$

$\left\{-5, \frac{7}{2}\right\}$

5. $\quad x(5-x) = 0 \quad 5 - x = 0$

$5 = x$

$x = 0$

$x = 5$

$\{0,5\}$

Quadratic Formula

The quadratic formula can help you solve quadratic equations that cannot be factored.

QUADRATIC FORMULA
$$x = \frac{-b \pm \sqrt{b^2 - 4ac}}{2a}$$

For example:
$x^2 - 6x = -3$
$x^2 - 6x + 3 = 0$

Can be solved using the following formula

Remember: $ax^2 + bx + c$
$a = 1$
$b = -6$
$c = 3$
Now you can plug it into the formula

$$x = \frac{-(-6) \pm \sqrt{(-6)^2 - 4(1)(3)}}{2(1)} = x = \frac{6 \pm \sqrt{36 - 12}}{2}$$

$$x = \frac{6 \pm \sqrt{24}}{2} = \frac{6 \pm \sqrt{(6)4}}{2} = \frac{6 \pm 2\sqrt{6}}{2} = 3 \pm \sqrt{6}$$

$$x_1 = 3 + \sqrt{6}$$

$$x_2 = 3 - \sqrt{6}$$

T

The quadratic formula can also help you to determine the amount of REAL solution of a quadratic equation can have.

> **DISCRIMINANT of QUADRATIC EQUATION**
> $$b^2 - 4ac$$
>
> If $b^2 - 4ac > 0$, There are 2 SOLUTIONS
> If $b^2 - 4ac = 0$, THERE IS ONE REAL 1 SOLUTION
> If $b^2 - 4ac < 0$, NO REAL SOLUTION

Let's Practice!

Determine the amount of solutions by using the discriminant, and then solve the equation by using the quadratic formula.

1. $x^2 + 11x + 30 = 0$
2. $x(x+2) = 5$
3. $2x^2 + x = 7$
4. $x^2 = -10 - 25$
5. $x^2 = x - 1$

Let's check your answers

1. $x^2 + 11x + 30 = 0$

   $a = 1$
   $b = 11$
   $c = 30$

   STEP ONE:

   $b^2 - 4ac$
   $(11)^2 - 4(1)(30)$
   $121 - 120 = 1$
   SINCE $b^2 - 4ac > 0$, THEN TWO REAL SOLUTIONS

   $$\frac{-b \pm \sqrt{b^2 - 4ac}}{2a} = \frac{-11 \pm \sqrt{(11)^2 - 4(1)(30)}}{2(1)}$$

   $\Rightarrow \dfrac{-11 \pm \sqrt{1}}{2} = \dfrac{-11 \pm 1}{2}$

$\Rightarrow \quad x_1 = \frac{-11-1}{2} = \frac{-12}{2} = -6 \qquad x_2 = \frac{-11+1}{2} = \frac{-10}{2} = -5$

2. $2x^2 + x = 7 \quad 2x^2 + x - 7 = 0 \quad \Rightarrow \quad a = 2, \ b = 1, \ c = -7$

   $b^2 - 4ac \Rightarrow (1)^2 - 4(2)(-7) = 1 + 56 = 57$

   Since $b^2 - 4ac > 0$ then **two real solutions**.

   $\frac{-b \pm \sqrt{b^2 - 4ac}}{2a} = \frac{-1 \pm \sqrt{(1)^2 - 4(2)(-7)}}{2(2)} = \frac{-1 \pm \sqrt{57}}{4} \quad x_1 = \frac{-1 + \sqrt{57}}{4}; \ x_2 = \frac{-1 - \sqrt{57}}{4}$

3. $x(x+2) = 5$

   $x^2 + 2x = 5$

   $x^2 + 2x - 5 = 0 \implies a = 1, \ b = 2, \ c = -5$

   $b^2 - 4ac = (2)^2 - 4(1)(-5) = 4 + 20 = 24$

   Since $b^2 - 4ac > 0$, then **two real solutions.**

   $\dfrac{-b \pm \sqrt{b^2-4ac}}{2a} = \dfrac{-2 \pm \sqrt{(2)^2-4(1)(-5)}}{2(1)}$

   $\dfrac{-2 \pm \sqrt{26}}{2} = \dfrac{-2 \pm \sqrt{6}\sqrt{4}}{2} = \dfrac{-2 \pm 2\sqrt{6}}{2}$

   $\implies x_1 = -1 + \sqrt{6} \quad \text{and} \quad x_2 = -1 - \sqrt{6}$

4. $x^2 = -10x - 25$

   $x^2 + 10x + 25 = 0 \implies a = 1, \ b = 10, \ c = 25$

   $b^2 - 4ac = (10)^2 - 4(1)(25) \implies 100 - 100 = 0$

   Since $b^2 - 4ac = 0$, then **only one solution**.

   $\dfrac{-b \pm \sqrt{b^2-4ac}}{2a} = \dfrac{-10 \pm \sqrt{(10)^2-4(1)(25)}}{2(1)} = \dfrac{-10 \pm \sqrt{0}}{2} = -\dfrac{10}{2} = -5$

   $x = -5$

5. $x^2 = x - 1$

   $x^2 - x + 1 = 0 \implies a = 1, \ b = -1, \ c = 1$

   $b^2 - 4ac \implies (-1)^2 - 4(1)(1) = 1 - 4 = -3$

   Since $b^2 - 4ac < 0$, then **No Real Solutions.**

Complex Numbers

Complex numbers are used when you have no real solutions.
You use the letter **i** to label them as imaginary.
For example, for a real number, let's say
$$\sqrt{9} = 3, -3$$
Now if you have a negative root, then you can use complex numbers
$$\sqrt{-9} = 3i, -3i$$
or
$$\sqrt{-16} = -4i, 4i$$
You can put an i representing complex numbers:

Remember:
$$i^2 = 1$$
$$\sqrt{-1} = i$$

For example:

a. $\sqrt{-32} = \sqrt{16} \cdot \sqrt{-2} = 4\sqrt{-2} = \mathbf{4i\sqrt{2}, -4i\sqrt{2}}$

b. $\sqrt{-81} = \mathbf{9i, -9i}$

c. $\sqrt{-72} = \sqrt{-8} \cdot \sqrt{9} = 3\sqrt{-8} = 3 \times 2i\sqrt{2} = \mathbf{6i\sqrt{2}, -6i\sqrt{2}}$

d. $\sqrt{-125} = \sqrt{25} \cdot \sqrt{-5} = 5 \cdot \sqrt{-5} = \mathbf{5i\sqrt{5}, -5i\sqrt{5}}$

<u>Let's practice</u>: Simplify the following radicals.

(A). $\sqrt{-36}$ (B). $\sqrt{-120}$
(C). $\sqrt{-99}$ (D). $\sqrt{-225}$

<u>Let's check your answers:</u>

(A). $\sqrt{-36} = 6i, -6i$

(B). $\sqrt{-120} = \sqrt{-4} \cdot \sqrt{30} = 2i\sqrt{30}, -2i\sqrt{30}$

(C). $\sqrt{-99} = \sqrt{9} \cdot \sqrt{-11} = -3i\sqrt{11}, 3i\sqrt{11}$

(D). $\sqrt{-225} = 15i, -15i$

Let's talk about $i, i^2, i^4 ...$

When i is raised to different powers this is what happens.
$$i^2 = -1,$$
$$i^3 = i^2 \cdot i = (-1) \cdot i = -i,$$
$$i^4 = i^2 \cdot i^2 = (-1)(-1) = 1$$

But what happens if you have i^{125} ?

Let's do it.

i^{125} step 1: take 125 and divide by 4

```
     31
   _____
 4 ) 125
     124
     ___
      1
```

Find the remainder

Step 2: Since the remainder **is 1**, then your answer **is** i

> If your remainder is
> 0 then your answer is 1
> 1 then your answer is i
> 2 then your answer is -1
> 3 then your answer is $-i$
> 4 then your answer is 1

As you can see it follows a pattern.

Simplify:
(A). i^{215} (B). $i^5(i^6 + i^7)$ (C). $i^{10}(i^{100})$
(D). i^{-25} (E). $-i^{+10}$

Let's check your answers

(A). $\frac{215}{4} = 53\frac{3}{4}$, the remainder = 3, $i^{215} = -i$

(B). $i^5(i^6 + i^7) = i^{5+6} + i^{5+7} = i^{11} + i^{12} \Rightarrow$ Now do i^{11} and i^12

The first step is to divide 11 by 4. $\frac{11}{4} = 2\frac{3}{4}$,   Remainder $R = 3$,

$$i^{11} = -i$$

$$\frac{12}{4} = 3 , \quad \text{Remainder } R = 0,$$

$$i^{12} = 1$$

$i^5(i^6 + i^7) = -i + 1 \quad or \quad 1 - i$

(C). $i^{10}(i^{100}) = i^{110} \Rightarrow \frac{110}{4} = 27\frac{1}{2}$, $R = 1 = i^{110} = \mathbf{1}$

(D). $i^{-25} = \frac{1}{i^{25}} \Rightarrow \frac{25}{4} = 6\frac{1}{4}$, $R = 1 \Rightarrow i = \frac{1}{i} = \boldsymbol{i^{-1}}$

(E). $-i^{+10} = -i^{10} \Rightarrow \frac{10}{4} = 2\frac{1}{2}$, $R = 1 \Rightarrow i = -i^{10} = \boldsymbol{-i}$

Operations of Complex Numbers

You can add, subtract, multiply and divide complex numbers.
Let's do some examples:
It's important to remember when you have a complex number, the notation is the following:

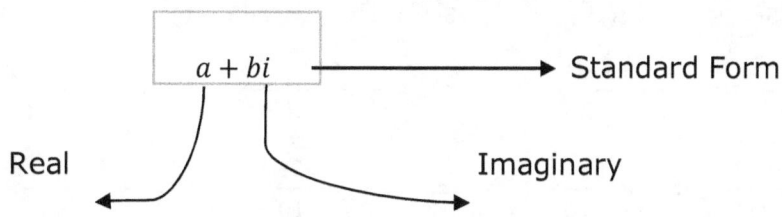

A). $(3 + 5i) - (2 + 6i) = 3 + 5i - 2 - 6i = \mathbf{1 - i}$

B). $(2i + 5)(3i - 6) = 6i^2 - 12i + 15i - 30 = \mathbf{6i^2 + 3i - 30}$
$6(-1) + 3i - 30 = -6 + 3i - 30 = \mathbf{-36 + 3i}$

C). $(2i + 3) + (6 - 8i) = 2i + 3 + 6 - 8i = \mathbf{9 - 6i}$

D). $2i \div 6 = \frac{2i}{6} = \frac{1}{3}i$

When you have a complex number, in the denominator you need to use the conjugate

> **Conjugate**
> If you have $a + bi$, then your conjugate is
> $a - bi$.
> If you have i, then your conjugate is -i

Let's do some examples:

A). $\dfrac{3}{i} = \dfrac{3}{i}\dfrac{(i)}{(i)} = \dfrac{3i}{i^2} = \dfrac{3i}{(-1)} = \boldsymbol{-3i}$

(B). $\dfrac{2}{-5i} = \dfrac{2(i)}{-5i(i)} = \dfrac{2i}{-5(i^2)} = \dfrac{2i}{-5(-1)} = \dfrac{2i}{5}$

(C). $\dfrac{2}{3+i} = \dfrac{2(3-i)}{(3+i)(3-i)} = \dfrac{6-2i}{9-3i+3i-i^2} = \dfrac{6-2i}{9-(-1)}$

$\dfrac{6-2i}{9+1} = \dfrac{6-2i}{10} = \dfrac{6}{10} - \dfrac{2i}{10} = \dfrac{3}{5} - \dfrac{1}{5}i$

Let's practice:
Simplify:

(A). $\dfrac{3-i}{2+5i}$ (B). $\dfrac{2-i}{1+i}$

(C). $\dfrac{8}{7i}$ (D). $\dfrac{-2+3i}{2-i}$

Let's check your answers:

(A). $\dfrac{1}{29} - \dfrac{17i}{29}$ (B). $\dfrac{1}{2} - \dfrac{3i}{2}$

(C). $\dfrac{-8i}{7}$ (D). $\dfrac{-7}{5} - \dfrac{4i}{5}$

Complex Solutions

Let's say you have the following quadratic equation:
$3X^2 + 5X = -7$, you can solve it using complex solutions.

Step 1 : Make the equation equal to zero

$$3X^2 + 5X + 7 = 0$$

Step 2 = Use the quadratic formula.

$$\frac{-b \pm \sqrt{b^2 - 4ac}}{2a} = \frac{-5 \pm \sqrt{(5)^2 - 4(3)(7)}}{2(3)} = \frac{-5 \pm \sqrt{25 - 84}}{6}$$

$$\frac{-5 \pm \sqrt{-59}}{6} = \frac{-5 \pm i\sqrt{59}}{6} = \frac{-5}{6} \pm \frac{i\sqrt{59}}{6}$$

As you can see, the procedure is the same but you now use the i to help you find the complex solutions.

Let's practice:

Find the roots (solutions):

(A). $X^2 + 1 = 0$ 　　　　　　　(B). $X^2 + 2X + 2 = 0$
(C). $-X^2 = 5X$ 　　　　　　　(D). $X^2 + 4 = 0$

Let's check your answers.

(A). $i, -i$ 　　　　　　　(B). $-1 + i, \; -1 - i$

(C). $\frac{5}{2} + \frac{\sqrt{11}i}{2}, \frac{5}{2} - \frac{\sqrt{11}i}{2}$ 　　　(D). $2i, -2i$

Polynomials

A polynomial is an expression of more than two algebraic expressions.
Let's say you have the following polynomial.

leading coefficient ← $5x^3$ — degree
$$5x^3 - 2x^2 + x - 2$$

This polynomial **has 4 terms**, its **degree is 3**(longest exponent).
This polynomial is **written in standard form**, that is, it is organized in a descending order with always the highest degree first.

For example;
Write the following polynomial in standard form:
$$-5x^2 + 6x^5 - 2x + 15 - 16x^3 + 10x^4$$
Standard form = $6x^5 + 10x^4 - 16x^3 - 5x^2 - 2x + 15$
The degree is 5 with 6 terms.

Let's do another example:
Write the following polynomial in standard form:

$6x^2 - 5x + 16x^3 - 1$ Since x^3 is the highest exponent, then
$16x^3 + 6x^2 - 5x - 1$ $16x^3$ is the leading exponent.
                              The degree is 3.
    → Standard form

The following chart shows the classification of polynomials by **name and terms.**

Term	Name	Example
1	Monomial	x^3
2	Binomial	$x^3 + 2$
3	Trinomial	$x^3 + 3x + 6$
4+	Polynomial	$x^3 + 2x^2 + x + 6$

Let's practice: Simplify the following polynomials in standard form:

a. $(-3x^5 + 6x - 7x^2) - (6x^2 - 5)$
b. $(2x - 3)(6y - 2x)$
c. $(x^6 - 3x^2)(2x - 9)$
d. $(2x^2 + x - 3) + (-5x^2 - 3x + 5)$

Let's check your answers:

a. $-3x^5 + 6x - 7x^2 - 6x^2 + 5 = -3x^5 - 13x^3 + 6x + 5$
degree = 5

b. $(2x - 3)(6y - 2x) = 12xy - 4x^3 - 18y + 6x^2$
$-4x^3 + 6x^2 + 12xy - 18y$ degree = 3

c. $(x^6 - 3x^2)(2x - 9) = 2x^7 - 9x^6 - 3x^3 + 27x^2$
degree = 7

d. $(2x^2 + x - 3) + (-5x^2 - 3x + 5) = 2x^2 + x - 3 - 5x^2 - 3x + 5$
$-3x^2 - 2x + 2$ degree = 2

Functions

A **function** is a relationship that assigns to each input number **EXACTLY ONE** output number.

> **If x is repeated more than one time, then the relationship is not a function.**

For example:

$$\{(2,0),(3,9),(1,9),(1,1)\}$$

Every **"x" is different**; therefore, this **relationship is a function.**

$$\{(2,0),(3,9),(2,9),(1,1)\}$$

There are two coordinates with the same value for "x" (2, 0) and (2, 9), this relationship is **NOT** a function

$F(x)$ or $G(x)$ or $H(x)$ is the same as "y". It's just a fancier notation and it is read as
"F of x", or "G of x".

For example, if we have a function:

F(x) = 3x^2 - 5 and we need F(2), that means that in every "x" you need to substitute it with a 2.

F(2) = 3(2)²-5
F(2) = 3(4) -5
F(2) = 12 -5 = 7

Therefore, when x= 2, then F(2)= 7 or y=7

Let's Practice!

If F(x)= 4x + 5. What is F(3)?

F(3) = 4(3) + 5
12 + 5= 17

When x=3, then y= 17 or

(3,17)

What is F(-1)?

F(-1) = 4(-1) + 5
-4 + 5= 1

When x= -1 then y= 1

(-1,1)

Sometimes they can ask you for the value of x if you have F(x).
For example:
If F(x) =3,
F(x) = 5x+18

Substitute F(x) = 3 and solve for x.

3= 5x+18
3-18 = 5x
-15= 5x
x=-3

(-3,3)

Let's do another example

If F(x)=2x-5, find F(4+a)
Substitute (4+a) into your x.
F(4+a) = 2(4+a)-5
F(4+a) = 8+2a-5
F(4+a) = 3+2a

Let's Practice:
If F(x) = $3x^2 - x + 5$, find:
1. F(-1)
2. F(2)
3. F(a)
4. If F(x)=4x-8, Find the value of x if F(x)=12
5. If F(x)=3x, Find the value of x if F(x)=-18

Let's check your answers
1) F(-1)= $3(-1)^2 - (-1) + 5$ =3(1)+1+5= 9
2) F(2)= $3(2)^2 - (2) + 5$ =3(4)-2+5= 15
3) F(a)= $3(a)^2 - (a) + 5$ =$3a^2 - a + 5$
4) 4x-8=12, 4x=12+8, 4x=20, x=5
5) -18=3x, x=-18/3= -6, x=-6

Symbolism

Let's say you have a function $f(x) = 3x + 2$

If you need to find $f(2)$, you just substitute the "2" inside the variable x.

$$f(x) = 3x + 2 \rightarrow f(2) = 3(2) + 2 = 6 + 2 = 8$$
$$f(2) = 8$$

On the test, you will have symbols that follow the same method.

Let's do an example.

If $x \spadesuit y = 2x + 3y$ then $1 \spadesuit 8$?

Just see the pattern $x \spadesuit y$ and $1 \spadesuit 8$

x is 1 and y is 8, just substitute

$$x \spadesuit y = 2x + 3y \rightarrow 1 \spadesuit 8 = 2(1) + 3(8) = 2 + 24 = 26$$

Let's practice with another example:

$a \odot b = \frac{a^b - b}{3a + b}$, what is the value of $2 \odot 4$.

Just substitute for $a = 2$ and $b = 4$.

$$2 \odot 4 = \frac{2^4 - 4}{3(2) + 4} = \frac{16 - 4}{6 + 4} = \frac{12}{10} = \frac{6}{5}$$

Let's practice:

    a.   If $x \triangle y = \frac{-x}{y^y}$, then find $-1 \triangle 2$.

    b.   If $b \ast y = 3b - 2^y$, then find $4 \ast 0$.

    c.   If $a \square b = \frac{2a + 3b}{-3b}$, then find $2 \square -1$.

    d.   If $x \odot y = \frac{(x+y)(x-y)}{x-y}$, then find $-5 \odot 2$.

Let's review your answers:

a. If $x \triangle y = \frac{-x}{y^y}$, then $-1 \triangle 2 = \frac{-(-1)}{2^2} = \frac{1}{4}$.

b. If $b \ast y = 3b - 2^y$, then $4 \ast 0 = 3(4) - 2^0 = 12 - 1 = 11$.

c. If $a \square b = \frac{2a+3b}{-3b}$, then $2 \square -1 = \frac{2(2)+3(-1)}{-3^{-1}} = \frac{4-3}{-3^{-1}} = \frac{1}{\frac{-1}{3}} = -3$.

d. If $x \odot y = \frac{(x+y)(x-y)x}{x^{-y}}$, then

$-5 \odot 2 = \frac{(-5+2)(-5-2)}{-5^{-2}} = \frac{(-3)(-7)}{-5^{-2}} = \frac{21}{\frac{-1}{25}} = (21)(-25) = -525$.

Composite Functions

$$(f \circ g)(X) = f(g(X))$$

1) f is eating g(x), you plug the function g(x) into f(x)
2) The small circle means "f composed of g".

If you have f(x)= 3x+2 and g(x)= $X^2 + 1$

Then f(g(x)), it is the function g(x) inside f(x)

Plug $g(X)$ into the X's of $f(X)$

$$(f \circ g)(X) = f(g(X)) = 3(X^2 + 1) + 2 = 3X^2 + 3 + 2 = \mathbf{3X^2 + 5}$$

Find:

$$g(f(X)) = (3X + 2)^2 + 1 = 9X^2 + 12X + 4 + 1 = \mathbf{9X^2 + 12X + 5}$$

Let's do another example:
Find
$$(f \circ g)(-1) = f(g(-1)) = 3((-1)^2 + 1) + 2 = 3(2) + 2 = 8$$
Plug $X = -1$ into $g(X)$
The result is then plug into $f(X)$.

Let's practice: If $f(X) = -2X^2 + 5$ $g(X) = 3X - 1$
(a). $f(g(X))$ (b). $g(f(-2))$
(c). $g(f(X))$ (d). $f(g(0))$

Let's check your answers:

(a). $f(g(X)) = -2(3X - 1)^2 + 5$
$= -2(9X^2 - 6X + 1) + 5 = -18X^2 + 12X - 2 + 5$
$= -18X^2 + 12X + 3$

(b). $g(f(-2)) = 3(-2X^2 + 5) - 1 = 3(-2(-2)^2 + 5) - 1 = 3(-2(4) + 5) - 1$
$= 3(-8 + 5) - 1 = 3(-3) - 1 = -10$

(c). $g(f(X)) = 3(-2X^2 + 5) - 1 = -6X^2 + 15 - 1 = -6X^2 + 14$

(d). $f(g(0)) = -2(3(0) - 1)^2 + 5 = -2(-1)^2 + 5 = 3$

How to Graph the Quadratic function

The **Parabola** is the graph of the quadratic function.

> The **standard form** of the quadratic function is:
> $$y = ax^2 + bx + c$$

For example, if you have a quadratic function like this one:
$$y = 2x^2 - 10x + 25,$$
then you can identify the letters a, b and c.
a= 2, b= -10 and c= 25.

These letters are important since they will be used to graph and understand the behavior of the parabola.

Let's say you need to graph $y = x^2$ and $y = -x^2$

When a>0, then the parabola is upward and you have a minimum point. It is easy to remember when a is positive the function is smiling ☺.

$y = 1x^2$, the value of a is 1. The parabola is happy so it looks **upward** (like a happy smile)253

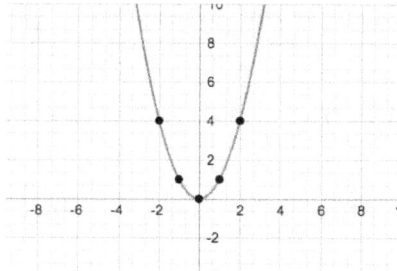

When a<0, then the parabola is downward and you have a Maximum point.

It is easy to remember when **a is negative** the function is upset ☹.

$Y = -1x^2$, the value of a is -1.

The parabola is upset so it looks **downward** (like an Unhappy smile)

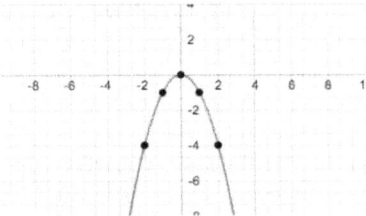

The **vertex** of the parabola helps you find the **minimum** or the **maximum** point.

> Vertex formula:
>
> To find the value of $x = \frac{-b}{2a}$
>
> To find the value of Y, once you get the value of x, then plug it into the original formula. $F\left(\frac{-b}{2a}\right)$.

Let's do some examples:

For the following parabola identify and **find the minimum or maximum point, then graph it.**

$$y = 2x^2 + 10x + 1$$

Let's first identify if we have a minimum or a maximum point

a) How can identify **if I have a minimum or maximum** point?

The value of a, b and c are:

a=2, b=10 and c= 25.

Since a= 2 is positive (she is happy) then we have a minimum point and my graph is upward.

b) How can I find the minimum point?

Let's use the formula to find the x-coordinate of the vertex.

$x = \frac{-b}{2a}$, **then using the letters a=2 and b=10,** then

$$x = \frac{-b}{2a} = \frac{-10}{2(2)} = -\frac{10}{4} = -2.5$$

Now that we know that x=-2.5, we can find the value of the y-coordinate by plugging -2.5 into the original function.

$$F(-2.5) = 2(-2.5)^2 + 10(-2.5) + 1 = -11.5$$

The vertex is located on (-2.5, -11.5)

We have a MINIMUN point on **-11.5.**

Domain and Range of QUADRATIC FUNCTIONS
The DOMAIN is ALWAYS all real numbers
$(-\infty, \infty)$
IF $a > 0$, The range is y≥ (the value of the minimum point)
IF a<0, The range is y≤ (the value of the maximum point)

CONCLUSION:

*The vertex is (-2.5, -11.5).

* The minimum point is -11.5.

* The domain is all real numbers. $(-\infty, \infty)$

* The range (since a>0) then y≥-11.5 or $[-11.5, \infty)$

* Using the function $y = 2x^2 + 10x + 1$, you can plug more values and graph it. Let's plug x= -1, x=0 and x= 1, then you have the points shown as black dots (-1,-7),(0,1) and (1,13)

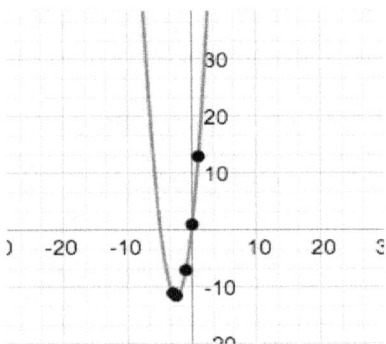

Let's practice:

Find the vertex, the minimum or maximum point, the domain and range and then graph the parabola.

1. $y = -5x^2 + 4x$
2. $y = 6x^2 - 1$
3. $y = \frac{1}{2}x^2 - 3x + 9$

Let's check your answers:

1. $y = -5x^2 + 4x$

   a=-5 b=4 c=0

   Using the formula to find the vertex:

   $x = \frac{-b}{2a} = \frac{-4}{2(-5)} = \frac{-4}{-10} = \frac{2}{5}$, find y by plugging into the formula:

$= -5(\frac{2}{5})^2 + 4(\frac{2}{5}) = 4/5$. **Vertex (2/5, 4/5)**

Since the value of **a is negative**, then the parabola has a **Maximum** point at **4/5.** The domain is all real numbers and the range is y≤4/5.

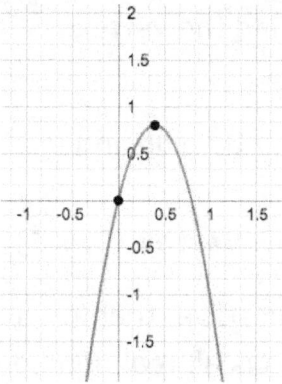

The graph shows the Maximum point at 4/5 or 0.75

2. $y = 6x^2 - 1$

a= 6 b=0 and c= -1

Using the formula to find the vertex:

$x = \frac{-b}{2a} = \frac{-0}{2(6)} = 0$, find y by plugging into the formula:

$= 6(0)^2 - 1 = -1$. **Vertex (0,-1)**

Since the value of a is positive, then the parabola has a **Minimum** point at **-1.** The domain is all real numbers and the range is y≥-1.

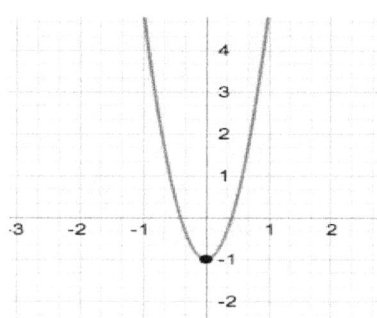

3. $y = \frac{1}{2}x^2 - 3x + 9$

a= $\frac{1}{2}$ b=-3 and c= 9

Using the formula to find the vertex:

$x = \frac{-b}{2a} = \frac{-(-3)}{2(1/2)} = \frac{3}{1} = 3$, find y by plugging into the formula:

$y = \frac{1}{2}(3)^2 - 3(3) + 9 = \frac{9}{2} = 4.5$ **Vertex (3,4.5)**

Since the value of a is positive, then the parabola has a **Minimum** point at **4.5.** The domain is all real numbers and the range is y≥4.5

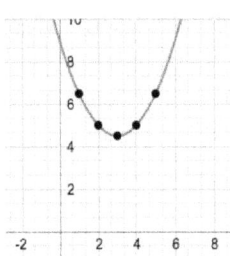

Transformation Rules

Transformation makes graphing easier if you follow these rules.

Function Notation	Type of Transformation	Movement of Graph	Change to Coordinate Point
$f(x) + k$	Vertical translation up k units	Shifts **up** k units	**Add** k to y $(x, y+k)$
$f(x) \pm k$	Vertical translation down k units	Shifts **down** k units	**Subtract** k from y $(x, y-k)$
$f(x + h)$	Horizontal translation left h units	Slides graph **left** h **units**	**Subtract** h from x $(x-h, y)$
$f(x - h)$	Horizontal translation right h units	Slides graph **right** h **units**	**Add** h to x $(x+h, y)$
$-f(x)$	Vertical reflection over $x-axis$	Flips graph over $x-axis$	Take the opposite value of y $(x, -y)$
$f(-x)$	Horizontal reflection over $y-axis$	Flips graph over $y-axis$	Take the opposite value of x $(-x, y)$
$af(x)$	Vertical stretch for $|a| > 0$	Pulls y values away from $x-axis$	Multiply y by a (x, ay)
$af(x)$	Vertical compression for $0 < |a| < 1$	Pushes y values toward $x-axis$	Multiply y by a (x, ay)

$f(bx)$	Horizontal compression for $|b|>0$	Pulls x-values away from $y-axis$	Divide x by b $\left(\frac{x}{b}, y\right)$
$f(bx)$	Horizontal stretch for $0<|b|<1$	Pushes x values towards $x-axis$	Divide x by b $\left(\frac{x}{b}, y\right)$

Symmetry of Functions

Functions can be even/odd or neither. Let me show you some examples.

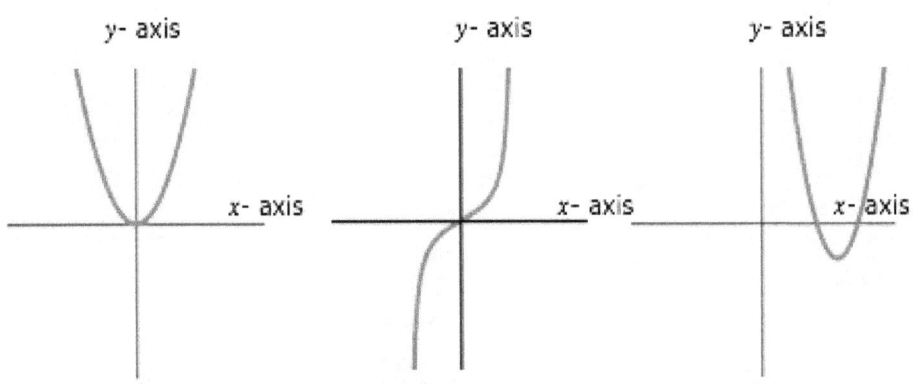

Even: Symmetric with the y-axis

Odd: Symmetric with the origin

Neither: No symmetry

Even Functions

Even functions are even symmetric with the y-axis.

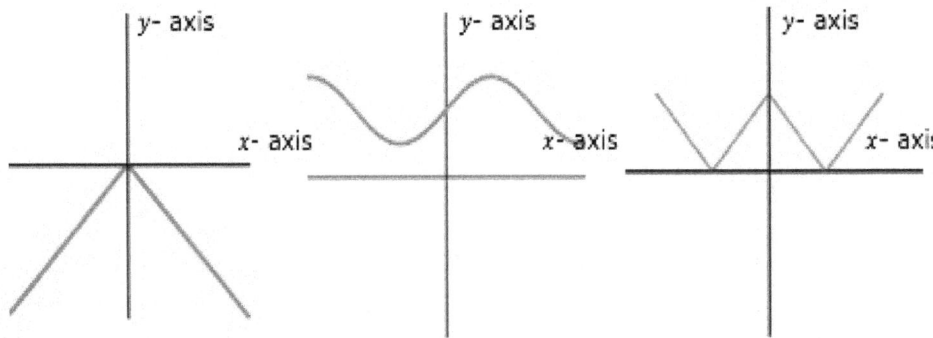

How to determine if a function is **even algebraically**?

For an even function **when you x and $-x$ into the function, you will get the same y value.**

For Example:
Let's say you have the function
$$f(x) = 3x^4 + 2x^2$$
Plug x and $-x$
$$f(x) = 3(x)^4 + 2(x)^2 = \mathbf{3x^4 + 2x^2}$$

$$f(-x) = 3(-x)^4 + 2(-x)^2 = \mathbf{3x^4 + 2x^2}$$

If you get the **same functions, then is an even function**

Determine if the following functions are even:

a. $f(x) = -3x^2 + 1$
$f(x) = -3(x)^2 + 1 = \mathbf{-3x^2 + 1}$

Same functions!

$f(-x) = -3(-x)^2 + 1 = \mathbf{-3x^2 + 1}$

Yes, even.

b. $f(x) = 2x^4 - 3x - 5$
$f(x) = 2(x)^4 - 3(x) - 5 = \mathbf{2x^4 - 3x - 5}$

Not the same functions!

$f(-x) = 2(-x)^4 - 3(-x) - 5 = \mathbf{2x^4 + 3x - 5}$

Not even!

Odd Functions:

Odd functions are symmetric with the origin, that is the point (0,0).

For example

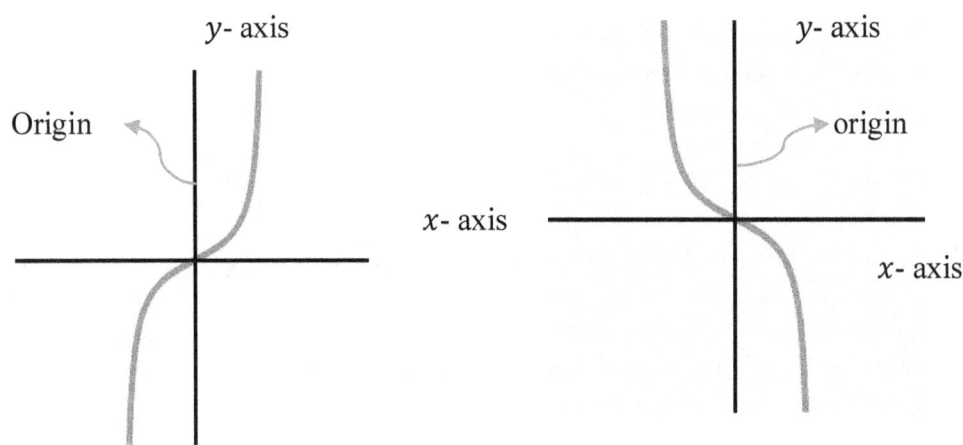

How to determine if a function is **odd algebraically**?

You need to do the following **two** steps:
1. Plug x into the function.
2. Plug $-x$ into the function and then change the signs.
   If you have the same functions, then the function is **odd**.

For example:

a. $f(x) = x^3 - 2x$

   1) Plug x into the function
      $$f(x) = (x)^3 - 2x = x^3 - 2x$$
   2) Plug $-x$ into the function and then change the signs
      $$f(x) = (-x)^3 - 2(-x) = -x^3 + 2x$$

   Change the signs = $x^3 - 2x$ it is an odd function

b. $f(x) = 4x^5 - 2x$

   1) Plug x into the function
      $$f(x) = 4(x)^5 - 2(x) = 4x^5 - 2x$$
   2) Plug $-x$ into the function and then change the signs

      $$f(x) = 4(-x)^5 - 2(-x) = -4x^5 + 2x$$

   Same function

   Change the signs = $4x^5 - 2x$

   it is an odd function

c. $f(x) = -2x^3 + x^2 - 1$

   1) Plug x into the function
      $$f(x) = -2(x)^3 + (x)^2 - 1 = -2x^3 + x^2 - 1$$
   2) Plug $-x$ into the function and then change the signs

      $$f(x) = -2(-x)^3 + (-x)^2 - 1 = 2x^3 + x^2 - 1$$

   Not Same function

   Change the signs = $-2x^3 - x^2 + 1$ it is **not** odd.

If a function is **even**, then it is **symmetric with the y- axis**.

(Like a parabola)

If a function **is odd**, then it is **symmetric with the origin** $(0,0)$

(Like a cube function)

If a function is **neither even nor odd, then the function is not symmetric.**

Let's practice: Determine if the following functions are even, odd or neither.

a. $f(x) = 3x^4 + 2x$
b. $f(x) = 2x^2 + x - 1$
c. $f(x) = -2x^3 + 2x^2$
d. $f(x) = -x^5 - x^3$

Let's check your answers:

a. $f(x) = 3x^4 + 2x$

Since the degree is 4(the highest exponent), the function is either even or neither

1) $f(x) = 3(x)^4 + 2(x) = 3x^4 + 2x$
2) $f(-x) = 3(-x)^4 + 2(-x) = 3x^4 - 2x$

Function is neither

b. $f(x) = 2x^2 + x - 1$

1) $f(x) = 2(x)^2 + (x) - 1 = 2x^2 + x - 1$
2) $f(-x) = 2(-x)^2 + (-x) - 1 = 2x^2 - x - 1$

Function is neither

c. $f(x) = -2x^3 + 2x^2$

Since the degree is 3, the function is either odd or neither

1) $f(x) = -2(x)^3 + 2(x)^2 = -2x^3 + 2x^2$

Not the same function,

2) $f(x) = -2(-x)^3 + 2(-x)^2 = 2x^3 + 2x^2$

then it is not odd.

Change the signs $= -2x^3 - 2x^2$

Function is neither.

d. $f(x) = -x^5 - x^3$

1) $f(x) = -(x)^5 - (x)^3 = -x^5 - x^3$

Same function, it is

2) $f(x) = -(-x)^5 - (-x)^3 = x^5 + x^3$

odd.

Change the signs $= -x^5 - x^3$

Exponential Equations

Let's say you have the following exponential equation.

$2^{X+2} = 32$ To find the value of X, you need to have the same base on both sides of the equation. In this case the base is 2.

$2^{X+2} = 2^5$

Let's remember $32 = 2^5$

$2^{X+2} = 2^5$ Having the same base 2, you eliminate them and Solve for X

$X + 2 = 5$ $X = 3$

Let's do another example:

Solve for the value of x

$$3^{X+7} = \frac{1}{27}$$

$\Rightarrow 3^{X+7} = \frac{1}{3^3} \Rightarrow 3^{X+7} = 3^{-3}$

(Remember the rules of exponent)

$X + 7 = -3$ $X = -10$

Let's Practice: Solve the following exponential equations.

(a). $4^{2X} \cdot 4^{-3X} = 16^{X+5}$

(b). $16^X \cdot 32^{X+2} = 1024$

(c). $Z^3 \cdot Z^{\frac{X}{2}+5} = Z^{10}$

(d). $49(7)^{3X-5} = 7^3$

Let's review your answers:

(a). $4^{2X} \cdot 4^{-3X} = 16^{X+5} \Rightarrow 4^{2X-3X} = 4^{2(X+5)} \Rightarrow 2X - 3X = 2(X+5)$

$-X = 2X + 10$ $-3X = 10$ $X = -\frac{10}{3}$

(b). $16^X \cdot 32^{X+2} = 1024 \implies 2^{4(X)} \cdot 2^{5(X+2)} = 2^{10} \implies 2^{4X+5(X+2)} = 2^{10}$
$4X + 5(X + 2) = 10 \qquad 4X + 5X + 10 = 10 \quad 9X = 0 \qquad X = 0$

(c). $Z^3 \cdot Z^{\frac{X}{2}+5} = Z^{10} \implies Z^{3+\left(\frac{X}{2}+5\right)} = Z^{10} \implies 3 + \frac{X}{2} + 5 = 10$

$8 + \frac{X}{2} = 10 \qquad \frac{X}{2} = 2 \qquad X = 4$

(d). $49(7)^{3X-5} = 7^3 \implies 7^2 \cdot 7^{3X-5} = 7^3 \qquad 7^{2+(3X-5)} = 7^3$
$2 + 3X - 5 = 3 \implies -3 + 3X = 3 \qquad 3X = 6 \; X = 2$

Logarithmic Functions

Logarithmic function is the inverse function of the exponential function. Let's go over the basic rules.

Convert the following from exponential into logarithmic expression.

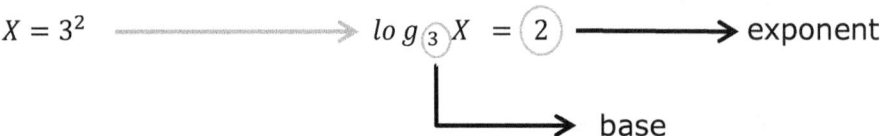

$X = 3^2 \longrightarrow \log_3 X = 2 \longrightarrow$ exponent

$\longrightarrow$ base

When you have a log without A base, by default the base is 10

Let's practice: Rewrite the following logs into the exponential form; and solve for the variable.

(a). $\log_8 X = 3$ (b). $\log_5 Y = -2$

(c). $\log_3 \frac{1}{27} = X$ (d). $\log 0.01 = X$

Let's check your answers:

(a). $\log_8 X = 8$ $8^3 = X$ $X = 512$

(b). $\log_5 Y = -2$ $5^{-2} = Y$ $Y = \frac{1}{5^2} = \frac{1}{25}$

(c). $\log_3 \frac{1}{27} = X$ $3^X = \frac{1}{27}$ $3^X = \frac{1}{3^3} \Rightarrow 3^X = 3^{-3}$

$X = -3$

(d). $\log(0.01) = X$ $\log_{10}(0.01) = X$ Remember if there is no base, by default will be 10

$$10^X = 0.01 \qquad 10^X = \frac{1}{100} \qquad 10^X = \frac{1}{10^2}$$

$$10^X = 10^{-2} \qquad X = -2$$

Rewrite the following exponential expression into a logs expression.

(a). $X^3 = 5$ \qquad (b). $2^{-X} = \frac{1}{32}$

(c). $5^3 = X$ \qquad (d). $10^{-X} = X$

Let's check your answers:

(a). $\log_X 5 = 3$

(b). $\log_2 \frac{1}{32} = -X$

(c). $\log_5 X = 3$

(d). $\log X = -3$

Properties of Logarithms

1. $\log_a(MN) = \log_a M + \log_a N$
2. $\log_a\left(\frac{M}{N}\right) = \log_a M - \log_a N$
3. $\log_a M^r = r \log_a M$
4. $\log_a M = \frac{\log M}{\log a} = \frac{\ln M}{\ln a}$

change *of* base formula:
$$\log_b a = \frac{\log a}{\log b}$$

Let's do some examples:

Write the following express as a single logarithm.

(a). $\log_b 10 + 3\log_b 5$ (Always start with the exponent rule)

   (1). $\log_b 10 + \log_b 5^3$ (Exponent rule)

   (2). $\log_b(10 \cdot 5^3)$ (Multiplication)

   (3). Final answer $\log_b(10) + 3\log_b 5 = \log_b(1250)$

Let's do another example:
Expand the following logarithm expression.

$\log_3\left(\frac{\sqrt[5]{X+5}}{X^2}\right)$ Since it is division you need to subtract

$\log_3 \sqrt[5]{X+5} - \log_3 X^2$ Now rewrite the radical

$\log_3(X+5)^{\frac{1}{5}} - \log_3 X^2$ and use the power rule

$\frac{1}{5}\log_3(X+5) - 2\log_3 X$ Final answer.

Let's practice:
Write each expression as a single logarithmic.

(a). $3\log_5 X + \frac{1}{2}\log_5 Y$

(b). $\frac{1}{3}\log_b X - 5\log_b Z$

(c). $3\ln a - \ln b^5$

(d). $5\log(X+1) + \frac{1}{5}\log Z$

Let's check your answers:

(a). $3\log_5 X + \frac{1}{2}\log_5 Y = \log_5 X^3 + \log_5 Y^{\frac{1}{2}} = \log_5 X^3 \cdot Y^{\frac{1}{2}}$
$= \log_5 X^3\sqrt{Y}$

(b). $\frac{1}{3}\log_b X - 5\log_b Z = \log_b(X)^{\frac{1}{3}} - \log_b Z^5 = \log_b \frac{X^{\frac{1}{3}}}{Z^5}$
$= \log_b \frac{\sqrt[3]{X}}{Z^5}$

(c). $\quad 3\ln a - \ln b^5 \;=\; \ln a^3 - \ln b^5 \;=\; \ln\dfrac{a^3}{b^5}$

(d). $\quad 5\log(X+1) + \dfrac{1}{5}\log Z \;=\; \log(X+1)^5 + \log Z^{\frac{1}{5}} \;=\; \log(X+1)^5 Z^{\frac{1}{5}}$

$\qquad\qquad\qquad\qquad\qquad\qquad\qquad\qquad\qquad\quad =\; \log(X+1)^5 \sqrt[5]{Z}$

Sequences

A patter or the repetition of numbers is called a sequence.
If the **difference is constant** between numbers,
then you have an **arithmetic sequence.**
Let's say you have the following arithmetic sequence.
$$-8, -3, 2, 7, 12, 17 \ldots a_n$$
a_n will be the term you need to find. a_1 is the first term,
d is the constant difference between two consecutive terms.

$-8, \quad -3, \quad 2, \quad 7, \quad 12, \quad 17 \ldots$ In this case $d = 5$ is the difference.

$a_1 \quad -3+5 \quad 2+5 \quad 7+5 \quad 12+5 \ldots$

You can also find d by subtracting any term by the term before.
For example:
$$d = -3 - (-8) = -3 + 8 = 5 \text{ or } \quad d = 7 - 2 = 5$$
The formula to find a_n:

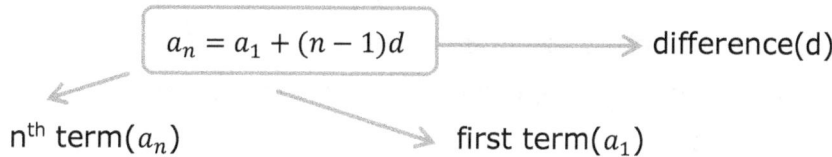

n^{th} term(a_n) first term(a_1)

Let's say you want a_{10} that is the 10th term.
$a_{10} = -8 + (10 - 1)(5) = -8 + 45 = 37$

1	2	3	4	5	6	7	8	9	10
-8	-3	2	7	12	17	22	27	32	37

<u>Let's practice:</u>   Identify if the following sequences are arithmetic and then find the 110th term.

(a). $-13, -5, 3, 11, 19, \ldots$ (b). $1, 3, 9, 18, 54, \ldots$

(c). $-23, -28, -33, -38, \ldots$ \hspace{2cm} (d). $9, 15, 21, 27, \ldots$

(e). $0, -3, -6, -9, -12, \ldots$

Let's check your answers:

(a). $\quad -13, -5, 3, 11, 19, \ldots$ the first step is to identify the variables of the formula.

$a_1 = -13$ \hspace{1cm} $d = 3 - (-5) = 3 + 5 = 8$ \hspace{1cm} $n = 110$

$a_n = a_1 + (n-1)d$ \hspace{1cm} $a_{110} = -13 + (110 - 1)(8)$

$a_{110} = 859$

(b). $\quad 1, 3, 9, 18, 54, \ldots$ The first term is $a_1 = 1$, $\quad d = 3 - 1 = 2$ but to be an arithmetic sequence all the difference must be constant, in this case d is not constant, then this is not an arithmetic sequence.

(c). $\quad -23, -28, -33, -38, \ldots$ \hspace{1cm} The first term is $a_1 = -23$

$d = -23 - (-23) = -28 + 23 = -5$, The difference is constant for all the terms, this is an arithmetic sequence

$a_n = a_1 + (n-1)d$ \hspace{1cm} $a_{110} = -23 + (110 - 1)(-5) = -568$

(d). $\quad 9, 15, 21, 27, \ldots$ The first term is $a_1 = 9$, $\quad d = 15 - 9 = 6$

The difference is constant for all the terms, this is an arithmetic sequence

$a_n = a_1 + (n-1)d$ \hspace{1cm} $a_{110} = 9 + (110 - 1)(6) = 663$

(e). $\quad 0, -3, -6, -9, -12, \ldots$ The first term is $a_1 = 0$, $\quad d = -3 - 0 = -3$

The different is constant for all the terms, this is an arithmetic sequence.

$a_n = a_1 + (n-1)d$ \hspace{1cm} $a_{110} = 0 + (110 - 1)(3) = -327$

Geometric Sequences

When your pattern has a **constant ratio**, then you have **a geometric sequence.**

For example: $-3, 6, -12, 24, -48, \ldots$

the ratio can be found by dividing any term by the term before.

$$R = \frac{6}{-3} = -2 \quad or \quad R = \frac{-48}{24} = -2$$

The formula to find the nth term:
$$a_n = a_1 \cdot R^{n-1}$$

Let's find the 20th term.
$$a_{20} = (-3)(-2)^{20-1} = (-3)(-2)^{19} = 1572864$$

Let's practice: Identify if the following sequences are geometric and then find the 10th term.

(a). $-5, -35, -245, -1715$

(b). $\frac{1}{3}, \frac{2}{3}, \frac{4}{3}, \frac{8}{3}, \ldots$

(c). $-3, -6, 12, -24, \ldots$

(d). $\frac{-3}{5}, \frac{-3}{10}, \frac{-3}{20}, \frac{-3}{40}, \ldots$

(e). $0.06, 0.36, 2.16, 12.96, \ldots$

Let's check your answers:

(a). $-5, -35, -245, -1715, \ldots$

The first term is -5. The ratio is $R = \frac{-35}{-5} = 7$

$a_n = a_1(R)^{n-1}$ $a_{10} = (-5)(-7)^{10-1} = (-5)(-7)^9 = -201768035$

(b). $\frac{1}{3}, \frac{2}{3}, \frac{4}{3}, \frac{8}{3}, ...$ The ratio is $R = \frac{\frac{2}{3}}{\frac{1}{2}} = 2$

$a_n = a_1(R)^{n-1}$ $a_{10} = \left(\frac{1}{3}\right)(2)^{10-1} = \left(\frac{1}{3}\right)(2)^9 = \frac{512}{3}$

(c). $-3, -6, 12, -24, ...$ The first term is -3. The ratio is $R = \frac{-6}{-3} = 2$

But the ratio is not constant. Then this is not a geometric sequence.

(d). $\frac{-3}{5}, \frac{-3}{10}, \frac{-3}{20}, \frac{-3}{40}, ...$ The first term $a_1 = \frac{-3}{5}$. The ratio $R = \frac{\frac{-3}{10}}{\frac{-3}{5}} = \frac{1}{2}$

$a_n = a_1 \cdot R^{n-1}$ $a_{10} = \left(\frac{-3}{5}\right)\left(\frac{1}{2}\right)^{10-1} = \left(\frac{-3}{5}\right)\left(\frac{1}{2}\right)^9 = \frac{-3}{2560}$

(e). $0.06, 0.36, 2.16, 12.96, ...$ The first term is 0.06

The ratio $R = \frac{0.36}{0.06} = 6$

$a_n = a_1 \cdot R^{n-1}$ $a_{10} = (0.06)(6)^{10-1} = (0.06)(6)^9 = 604661.71$

Series (adding sequences)

Let's say you want to add either an arithmetic or geometric sequences.
The following formulas are used to find arithmetic/geometric series.

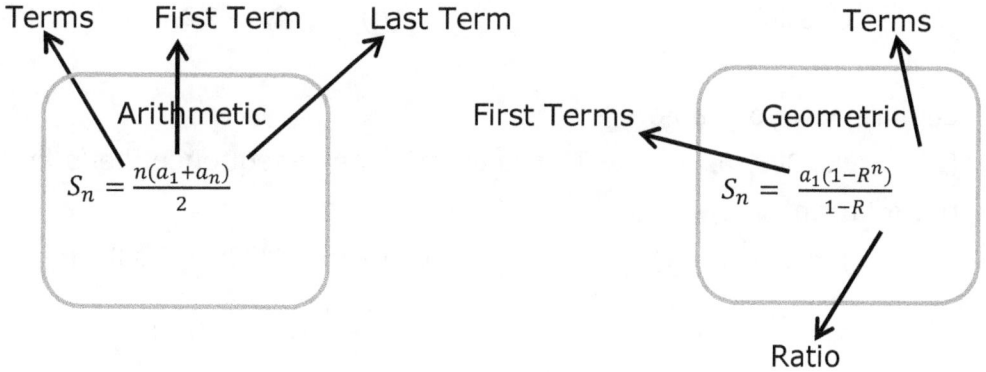

Let's do an example:
Find the sum of the first 30 terms of the following sequence.
2, 6, 10, 14, ...   The first step is to find the last term

$a_{30} = 2 + (30 - 1)(4) = 2 + 29(4) = 118$

Now you can use the formula

$S_n = \frac{n(a_1 + a_n)}{2}$ $S_{30} = \frac{30(2+118)}{2} = 1800$

Let's do another example:
Find the sum of the first 10 terms of the following sequence.

$-1, \frac{-5}{3}, \frac{-25}{9}, \frac{-125}{27}, \ldots$ The first step is to identify this is geometric

sequence since the ratio "R" is constant. $R = \frac{\frac{-5}{3}}{-1} = \frac{5}{3}$

$$S_n = \frac{a_1(1-R^n)}{1-R} = \frac{-1(1-\left(\frac{5}{3}\right)^{10})}{1-\frac{5}{3}} = \frac{-1(1-\left(\frac{5}{3}\right)^{10})}{\frac{-2}{3}} = -246.57$$

Let's Practice:

Find the sum of the first 6 terms of the following arithmetic and geometric sequences.

(a). 600, 300, 0, ...

(b). $-2, -1, -0.5, -\frac{1}{4}, ...$

(c). 10, 15, 20, 25, ...

(d). $-2, 2, -2, 2, ...$

Let's review your answers:

(a). 600, 300, 0, ... This is an arithmetic sequence. Let's find the 6th term

$$a_6 = a_1 + (n-1)d \qquad a_6 = 600 + (6-1)300 = 2100$$

$$S_n = \frac{n(a_1+a_n)}{2} \qquad S_6 = \frac{6(600+2100)}{2} = 8100$$

(b). $-2, -1, -0.5, -\frac{1}{4}, ...$ This is a geometric sequence.

Let's find ration $R = \frac{-1}{-2} = \frac{1}{2}$

$$S_n = \frac{a_1(1-R^n)}{1-R} = \frac{-2(1-\left(\frac{1}{2}\right)^6)}{1-\frac{1}{2}} = \frac{-2(1-\frac{1}{64})}{1-\frac{1}{2}} = \frac{-63}{16}$$

(c). 10, 15, 20, 25, ... This is an arithmetic sequence. Let's find the 6th term.

$$a_n = a_1 + (n-1)d = a_6 = 10 + (6-1)(5) = 10 + 25 = 35$$

$$S_n = \frac{n(a_1+a_n)}{2} = S_6 = \frac{6(10+35)}{2} = 135$$

(d). $-2, 2, -2, 2, ...$ This is a geometric sequence. Let's find the ratio

$R = \frac{2}{-2} = -1$

$S_n = \frac{a_1(1-R^n)}{1-R} = \frac{-2(1-(-1)^6)}{1-(-1)} = \frac{-2(1-1)}{2} = 0$

Sigma Notation

Series can also be noted using the sigma notation.
Let's do an example.

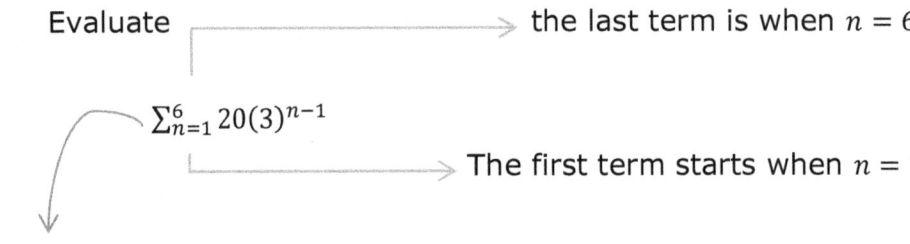

Sigma (means sum)

You can evaluate it by plugging the values.

$$\sum_{n=1}^{6} 20(3)^{n-1} = 20(3)^{1-1} + 20(3)^{2-1} + 20(3)^{3-1} + 20(3)^{4-1}$$

$$20(3)^{5-1} + 20(3)^{6-1}$$
$$= 20 + 60 + 180 + 540 + 1620 + 4860 = 7280$$

Let's do it using the formula. This is a geometric sequence, the $R = 3$

$$S_n = \frac{a_1(1-R^n)}{1-R} \qquad a_1 = 20, \qquad a_6 = 4860, \qquad n = 6$$

$$S_6 = \frac{20(1-(3)^6)}{1-3} = 7280$$

Let's do another example

$\sum_{n=1}^{13} 2n + 5$ This is an arithmetic sequence. Let's find a_1 and a_{13} to apply the formula

$a_1 = 2(1) + 5 = 7$ $a_{13} = 2(13) + 5 = 31$ $n = 13$

$$S_n = \frac{n(a_1+a_n)}{2}$$

$$S_{13} = \frac{13(a_1+a_{13})}{2} = \frac{13(7+31)}{2} = 247$$

Let's practice:

(a). $\sum_{n=2}^{20} -n + 5$ 　　　　　(b). $\sum_{n=1}^{15} -2(-3)^{n-1}$

(c). $\sum_{n=1}^{90} 3n$ 　　　　　(d). $\sum_{n=2}^{10} (-1)\left(\frac{1}{2}\right)^{n-1}$

Let's check your answers:

(a). -114 　　　(b). -7174454 　　　(c). 12285

(d). $\frac{-511}{512}$

Convergent and Divergent

When you have a geometric series and the ratio is between 0 and 1, then you have a convergent sequence.

For example:

$1, \frac{1}{2}, \frac{1}{4}, \frac{1}{8}, \frac{1}{16}\ldots\ldots$ the terms get smaller closer to "zero", they converge. The sum will yield to the following formula.

$$S = \frac{a_1}{1-R}. \quad \text{In our case}$$

$$S = \frac{1}{1-\frac{1}{2}} = \frac{1}{\frac{1}{2}} = 2$$

Let's say the ratio of a sequence is not between 0 and 1, then the series diverges.

If $0 < |R| < 1$	If the series do
Sum converges	not converge the
You need to use	diverges and you
Formula $S = \frac{a_1}{1-R}$	cannot use formula.
	The sum will be
	$-\infty$ or ∞

Let's practice: Evaluate the sum for the following geometric series.

(1). $\sum_{n=1}^{\infty} 4 \cdot \left(\frac{1}{2}\right)^{n-1}$

(2). $\sum_{n=1}^{\infty} 3 \cdot \left(\frac{9}{2}\right)^{n-1}$

(3). $\sum_{n=1}^{\infty} -2\left(\frac{-3}{4}\right)^{n-1}$

(4). $\sum_{n=1}^{\infty} 2\left(\frac{17}{10}\right)^{n-1}$

Let's check your answers:

(1). Since $R = \frac{1}{2}$, the series converges, $S = \frac{a_1}{1-R} = \frac{4}{1-\frac{1}{2}} = 8$

(2). Since $R = \frac{9}{2}$, the series diverges to ∞.

(3). Since $R = \frac{-3}{4}$, the series converges, $S = \frac{a_1}{1-R} = \frac{-2}{1-\left(\frac{-3}{4}\right)} = \frac{-8}{7}$

(4). Since $R = \frac{17}{10}$, the series diverges to ∞.

Compounding Interest

The following formula finds the future value of a lump sum.

$$A = P\left(1 + \frac{r}{n}\right)^{n \cdot t}$$

r = annual rate

- Compound period
- Principal (the amount you have today)
- Accumulate (Future Value)

$$A = P \cdot e^{r \cdot t}$$

Continuously Compounded

If the problem says:
- Yearly: $n = 1$
- Monthly: $n = 12$
- Quarterly: $n = 4$
- Semi-annual: $n = 2$
- Daily: $n = 365$
- Weekly: $n = 52$

Let's say you have $100 and deposit it into an account that yields 15% annualy. What is the future value in 20 years. If the account compounds;

(a). Yearly   (b). Monthly   (c). Quarterly
(d). Semi-annual (e). Continuously

Yearly	Monthly	Quarterly
$n = 1$ $$A = 100\left(1 + \frac{0.15}{1}\right)^{20 \cdot 1}$$ $1,636.65	$n = 12$ $$A = 100\left(1 + \frac{0.15}{12}\right)^{20 \cdot 12}$$ $1,971.55	$n = 4$ $$A = 100\left(1 + \frac{0.15}{4}\right)^{20 \cdot 4}$$ $1901.29

Semi-annual	Continuously
$n = 2$ $$A = 100\left(1 + \frac{0.15}{2}\right)^{20(2)}$$ $1804.42	$$A = 100 \cdot e^{0.15(20)}$$ $2,008.55

Let's practice:

Peter deposits $200 dollars in an account for 5 years at a rate of 18%
What is the future value if the account compounds:

(a). Yearly (b). Monthly (c). Quarterly

(d). Semi-annual (e). Continuously

Let's check your answer:

(a). $A = P\left(1+\frac{r}{n}\right)^{n \cdot t} = 200\left(1+\frac{0.18}{1}\right)^{5\times 1} = 457.55$

(b). $A = P\left(1+\frac{r}{n}\right)^{n \cdot t} = 200\left(1+\frac{0.18}{12}\right)^{5\times 12} = 488.64$

(c). $A = P\left(1+\frac{r}{n}\right)^{n \cdot t} = 200\left(1+\frac{0.18}{4}\right)^{5\times 4} = 482.34$

(d). $A = P\left(1+\frac{r}{n}\right)^{n \cdot t} = 200\left(1+\frac{0.18}{2}\right)^{5\times 2} = 473.47$

(e). $A = Pe^{n \cdot t} = 200 \times e^{5(0.18)} = 491.92$

The Circle

> **Standard form:**
> $(x - h)^2 + (y - k)^2 = r^2$
> **General form:**
> $Ax^2 + Ay^2 + Bx + Cy + D = 0$
> (h, k) is the center

You have the center of (−3,5) and radius of 6 for a circle.
Let's graph it and find its **standard and general form**

1. Plot the center

    (−3,5)

    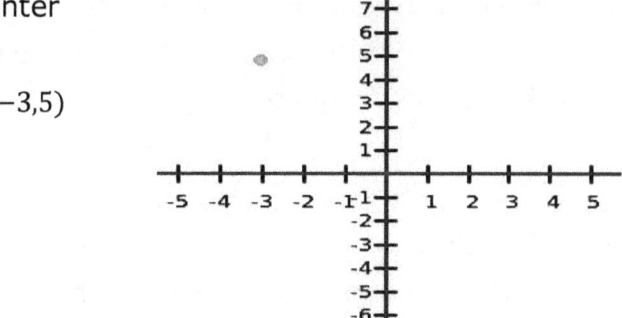

2. From center you can go up, down, right and left 6 units.

    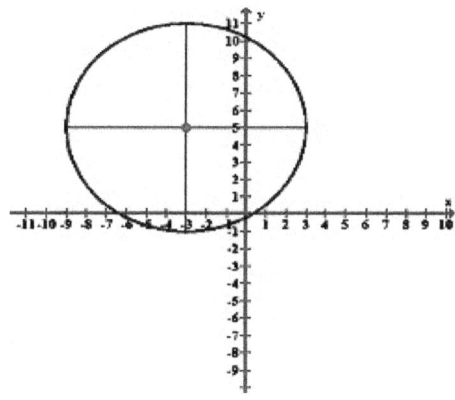

3. **Find the standard equation of the circle.**
   Remember, the formula is $(x-h)^2 + (y-k)^2 = r^2$,
   where h is value of x and k is value of y of center and r is radius.
   Then, $(x-(-3))^2 + (y-5)^2 = 6^2$
   $(x+3)^2 + (y-5)^2 = 36$ ⟶ Standard Equation

If you start expanding the parenthesis, then you have
$(x+3)(x+3) + (y-5)(y-5) = 36$
$x^2 + 3x + 3x + 9 + y^2 - 5y - 5y + 25 = 36$
$x^2 + 6x + 9 + y^2 - 10y + 25 = 36$

The final step is to put all constants together.

$x^2 + 6x + y^2 - 10y = 36 - 25 - 9$
$x^2 + 6x + y^2 - 10y = 2$ ⟶ Make equal to zero.
$x^2 + 6x + y^2 - 10y - 2 = 0$ ⟶ This is the **general form**.

Let's practice:
Find the standard and general form of the following circles :
a. $R = 10, C(2,5)$ b. $R = 8, C(-8,10)$

c. $R = \sqrt{6}, C(9,1)$ d. $R = 1, C(-2,-5)$

Let's check your answers:

a. $(x-h)^2 + (y-k)^2 = r^2$, then
$(x-2)^2 + (y-5)^2 = 10^2$ is the standard form.
$(x-2)(x-2) + (y-5)(y-5) = 100$
$x^2 - 2x - 2x + 4 + y^2 - 5y - 5y + 25 = 100$
$x^2 - 4x + 4 + y^2 - 10y + 25 - 100 = 0$
$x^2 - 4x + y^2 - 10y - 71 = 0$
$\longrightarrow$ general form

b. $(x-(-8))^2 + (y-10)^2 = 8^2$
$(x+8)^2 + (y-10)^2 = 64$ is the standard form.
$(x+8)(x+8) + (y-10)(y-10) = 64$
$x^2 + 8x + 8x + 64 + y^2 - 10y - 10y + 100 = 64$
$x^2 + 16x + 64 + y^2 - 20y + 100 - 64 = 0$
$x^2 + 16x + y^2 - 20y + 100 = 0$ general form

c. $(x-9)^2 + (y-1)^2 = (\sqrt{6})^2$
$(x-9)^2 + (y-1)^2 = 6$ $\longrightarrow$ standard form.
$(x-9)(x-9) + (y-1)(y-1) = 6$
$x^2 - 9x - 9x + 81 + y^2 - y - y + 1 = 6$
$x^2 - 18x + y^2 - 2y + 81 + 1 - 6 = 0$
$x^2 - 18x + y^2 - 2y + 76 = 0$ general form

d. $(x+2)^2 + (y+5)^2 = 1^2$
$(x+2)^2 + (y+5)^2 = 1$ ⟶ standard form.
$(x+2)(x+2) + (y+5)(y+5) = 1$
$x^2 + 2x + 2x + 4 + y^2 + 5y + 5y + 25 = 1$
$x^2 + 4x + y^2 + 10y + 4 + 25 - 1 = 0$
$x^2 + 4x + y^2 + 10y + 28 = 0$

⟶ general form

Exponential Function

We are only dealing with basic exponential functions now in algebra 1, later in algebra 2 you will have more fun with more complex ones. Let's go over the basics of exponential functions. The exponential function follows the following formula, it is important to understand its components.

$$f(x) = a(b)^x$$

WHERE:

a is the initial value

b is the rate of growth

or

decay (if b is a number less between 0 and 1)

X is the variable

Alert!
If $b > 0$ the function is increasing (growing really fast)

For example:

Let's say you have the following exponential function:

$$f(x) = 5(2)^x$$

Let's plug the following values for x= -2,-1,0,1,2 and graph it.

$$f(x) = 5(2)^x = 5(2)^{-2} = \frac{5}{2^2} = \frac{5}{4} = 1.25$$

$$f(x) = 5(2)^x = 5(2)^{-1} = \frac{5}{2^1} = \frac{5}{2} = 2.5$$

$$f(x) = 5(2)^x = 5(2)^0 = 5(1) = 5$$

$$f(x) = 5(2)^x = 5(2)^1 = 10$$

$$f(x) = 5(2)^x = 5(2)^2 = 5(4) = 20$$

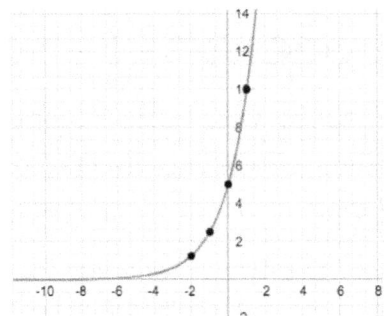

Notice how the y-value increases super fast!
What is the domain?
ALL REAL NUMBERS
What is the range?
y>0
What is the Y-intecept?
When x=0, then y is 5, (0,5)

Let's do another exponential function:
$$f(x) = 5(1/2)^x$$
Let's plug the following values for x= -2,-1,0,1,2 and graph it.

$$f(x) = 5(1/2)^x = 5(1/2)^{-2} = \frac{5}{(\frac{1}{2})^2} = \frac{5}{1/4} = 20$$

$$f(x) = 5(1/2)^x = 5(1/2)^{-1} = \frac{5}{(\frac{1}{2})^1} = \frac{5}{1/2} = 10$$

$$f(x) = 5(1/2)^x = 5(1/2)^0 = 5(1) = 5$$

$$f(x) = 5(1/2)^x = 5(1/2)^1 = 2.5$$

$$f(x) = 5(1/2)^x = 5(1/2)^2 = 5\left(\frac{1}{4}\right) = 1.25$$

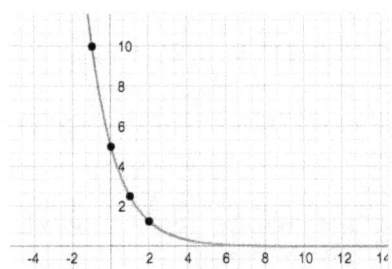

Notice how the y-value decreases super fast!

What is the domain?
ALL REAL NUMBERS
What is the range?
y>0
What is the Y-intecept?
When x=0, then y is 5, (0,5)

Alert!
The domain of exponential function is ALWAYS all real numbers $(-\infty, \infty)$
The y-intecept is found when x=0

Let's practice:

Graph the following functions using the values of x: -2,-1,0,1

a) $f(x) = 2(5)^x$

b) $f(x) = 4(3)^x$

c) $f(x) = 6\left(\frac{1}{3}\right)^x$

d) $f(x) = (0.625)^x$

Let's check your answers:

a) $f(x) = 2(5)^x$

The points are: (-2,0.08),(-1,0.4),(0,2),(1,10)

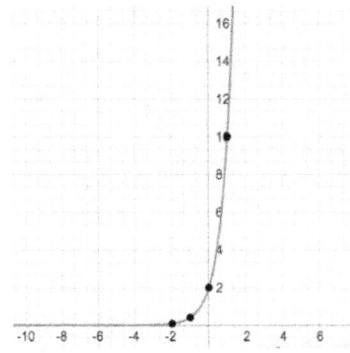

1) The domain is ALL REAL NUMBERS
2) The range is y>0
2) The y-intercept is (0,2)

b) $f(x) = 4(3)^x$

The points are: (-2,0.44),(-1,1.33),(0,4),(1,12)

Notice how the function is increasing

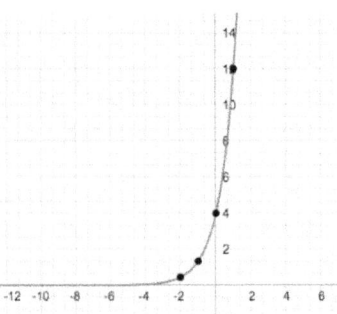

1) The domain is ALL REAL NUMBERS
2) The range is y>0
3) The y-intercept is (0,4)

c) $f(x) = 6\left(\dfrac{1}{3}\right)^x$

The points are: (-2,54),(-1,18),(0,6),(1,2)
Notice how the function is decreasing

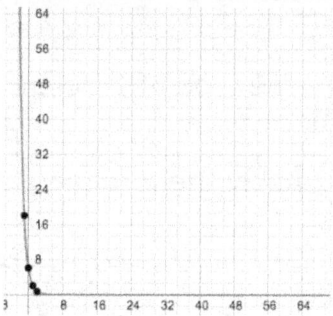

1) The domain is ALL REAL NUMBERS
2) The range is y>0
3) The y-intercept is (0,6)

d) $f(x) = (0.625)^x$

The points are: (-2,2.56),(-1,1.6),(0,1),(1,0.625)

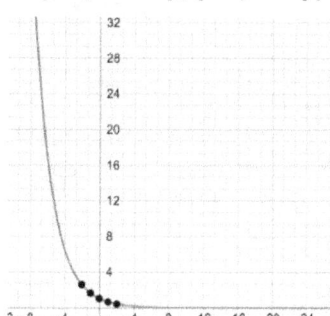

1) The domain is ALL REAL NUMBERS
2) The range is y>0
3) The y-intercept is (0,1)

Word Problems

1) Read the problem first

2) Identify the type of problem type

3) Apply the formula

Motion Problems

Rate, time and distance
$$d = r \cdot t$$

Average Rate
$$Average\ Rate = \frac{Total\ Distance}{Total\ Time}$$

Case I: Two cars originating from the same point but at different times, going in the same direction make the distances equal: d1=d2

A) Car "X" leaves point A at 9:00 am at a rate of 60 mph. Car "Y" leaves point A at 11 am at a rate of 90 mph. How long until both cars meet?

Solution: Both cases have different rates but they have <u>same</u> distance

$$d_1 = d_2$$
$$d_1 = t \cdot 60$$
$$d_2 = (t - 2) \cdot 90$$
$$t \cdot 60 = (t - 2)90$$
$$60t = 90t - 180$$
$$180 = 30t$$
$$\mathbf{t = 6 \text{ hours}}$$

b) Car "X" left Miami to New York. 6 hours later car "Y" left travelling at 70 mph to catch up with car X. After 4 hours both cars were in the same spot.
What was the average speed for car X?

Solution: Both cars have the same distance.
The distance is $d = R \cdot t = 70(4) = 280$ miles.
Since $d_1 = d_2$, we know that $d_2 = (6 + 4) \cdot X$
$$280 = 10X$$
$$\mathbf{X = 28 \text{ mph}}$$

Case II: Two cars are going away from each other in opposite directions

$$D_1 + D_2 = D_{Total}$$

a) Two cars leave the town at the same time going opposite direction.
One of them travels at 45 mph and the other at 55 mph.
In how many hours will they be 150 miles apart?

Solution:

$$d_1 = 45t \quad d_2 = 55t$$
$$d1 + d2 = d_T = 150$$
$$45t + 55t = 150$$
$$100t = 150$$
$$t = 1.5 \text{ hours}$$

Case III Two cars are going towards each other

$$D_1 + D_2 = D_{Total}$$

b) Two cars, travelling towards each other, left two towns that are 800 miles apart. If the rate of the first car is 70 mph and the second car is 30 mph, how long until they pass each other?

Solution:

$$D_1 = 70t \quad D_2 = 30t$$
$$D_1 + D_2 = D_T$$
$$70t + 30t = 800$$
$$100t = 800$$
$$t = 8 \text{ hours}$$

Rate/Work Problems

These problems are usually people working, a machine, or a water pipe.

$$\frac{1}{W_1} + \frac{1}{W_2} \ldots + \frac{1}{W_n} = \frac{1}{Total}$$

a) Machine A takes 3 hours to print labels. Machine B takes 5 hours doing the same job. How long will it take both machines if they work together?

Solution: $\quad \frac{1}{W_1} + \frac{1}{W_2} = \frac{1}{Total} \qquad \frac{1}{3} + \frac{1}{5} = \frac{1}{T} \qquad \frac{5+3}{15} = \frac{1}{T}$

$$T = \frac{15}{8} = 1\frac{7}{8} \ hours$$

b) Working together both machines can print labels in 6 hours. If machine A works twice as fast than machine B. How long does it take to do the job for machine A?

Solution: $\quad \frac{1}{0.5t} + \frac{1}{t} = \frac{1}{6} \qquad \diamond$ **Faster means $\frac{1}{2}$ of the time**

$\qquad \qquad \frac{1}{0.5t} + \frac{0.5}{0.5t} = \frac{1}{6} \qquad \qquad \frac{1.5}{0.5t} = \frac{1}{6}$

$\qquad \qquad 6(1.5) = 0.5t \qquad \qquad \qquad 9 = 0.5t$

$$t = \frac{9}{0.5} = 18 \text{ hours}$$

Machine A $= 0.5(18) = $ **9 hours**

Consecutive Integers

The first integer is always X, then if it is consecutive the next one is X+1,
but if it is consecutive even or odd, then second integer is X+2

> Consecutive $= X + (X + 1) + (X + 2) + \cdots$
> Even / Odd $= X + (X + 2) + (X + 4) + \ldots$

(a). The sum of four consecutive integers is 46. Find the integers.

$$X + (X + 1) + (X + 2) + (X + 3) = 46$$
$$4X + 6 = 40$$
$$4X = 40$$
$$X = 10$$

1st=10   **2nd=11**   **3rd=12**   **4th=13**

(b). The average of three odd consecutive integers is 15. Find the integers.

$$\frac{X + (X + 2) + (X + 4)}{3} = 15$$
$$X + (X + 2) + (X + 4) = 45$$
$$3X + 6 = 45$$
$$3X = 39$$
$$X = 13$$

    1ˢᵗ=13   **2ⁿᵈ=15**   **3ʳᵈ=17**

(c). The sum of four even integers is 20. Find the integers.

$$X + (X+1) + (X+2) + (X+4) = 20$$
$$4X + 12 = 20$$
$$4X = 8$$
$$X = 2$$

  1ˢᵗ=2   **2ⁿᵈ=4**   **3ʳᵈ=6**   **4ᵗʰ=8**

Mixture Problems

$$p_1(Volume_1) + p_2(Volume_2) = p_{mixture}(V_1 + V_2)$$

A) How many liters of a solution that is 12% salt must be added to 6 liters of a solution that is 8% so that the resulting solution is 10% salt?

$$p_1(Volume_1) + p_2(Volume_2) = p_{mixture}(V_1 + V_2)$$
$$0.12(V) + 0.08(6) = 0.10(V + 6)$$
$$0.12V + 0.48 = 0.10V + 0.6$$
$$0.12V - 0.10V = 0.6 - 0.48$$
$$0.02V = 0.12$$
$$V = 6 \; liters$$

b) Vera mixed together 9 gallons of Brand "A" fruit drink and 8 gallons of Brand "B" fruit drink which contains 48% fruit juice. If the mixture contained 30% fruit juice, what is the fruit juice concentration of Brand A?

$$p_1(Volume_1) + p_2(Volume_2) = p_{mixture}(V_1 + V_2)$$
$$p_1(9) + 0.48(8) = 0.30(9 + 8)$$
$$9p_1 + 4.32 = 0.30(17)$$
$$9p_1 = 5.1 - 3.84$$
$$p_1 = 14$$

Concentration of Brand A is 14%

Simple Interest Problems

The simple interest problems follow this formula.

$$I = P \cdot R \cdot t$$
p is principal
r is the annual rate
t is the time

(a). If $5,000 is invested at 10% annual interest. How much interest is earned after 8 months?

$$I = 5000(0.10)\left(\frac{8}{12}\right) = 333.33$$

$$I = \$333.33$$

b) Lupe invested $6000 in a certificated deposit for 4 years. She earned a total of $480 in interest, what was the annual rate?

$$480 = 6000(r)(4)$$
$$r = (480)/24000 = 0.02$$
r=2%

Profit Problems

$$Revenue = Price * quantities$$
$$Cost = Cost * quantities$$

$$Profit = Revenue - Cost$$

Revenue > Cost is a Profit
Revenue < Cost is a Loss

(a). The revenue function of a company is $R(X) = 7.5X$ and its cost functions is $C(X) = 2.5X + 1$. Find the profit or loss if the company sells 20 units.

$$P(X) = R(X) - C(X) = 7.5X - (2.5X + 1)$$
$$P(X) = 7.5X - 2.5X - 1 = 5X - 1$$
$$P(20) = 5(20) - 1 = 100 - 1 = \$\,99$$

The company has a profit of $99

b) Empire Records sells its CD's at $4 dollars each. The cost of each CD is $1.50 plus a fixed cost of $100. How many CD's will they have to sell to break even?

The breakeven point is when the Revenue is equal to the cost

Revenue is 4X, the cost is 1.5X+ 100

4x=1.5x+100

$$2.5X = 100$$
$$x = 100/2.5 = 40 \text{ CDs}$$

Money Problems

There will be questions with money and number of items, the general formula is the one shown below. Usually these questions involve candy, pencils, notebooks or other items you buy at a store.

$$\text{item } 1 + \text{item } 2 = \text{Total items}$$

a) Kristine went to the candy store and bought 22 candies Brand A and B spending $6 in total. If Brand A costs $0.30 and Brand B costs $0.25, how many candies of brand B did she buy?

1) How many candies did she buy?

A + B = 22

2) How much did she spend in total?

$0.30A + $0.25B = $6

Let's solve for A:
A = 22 - B

Substitute into the second equation:
0.30(22-B) + 0.25B = 6

$6.6 - 0.30B + 0.25B = 6$

$-0.005B = 6 - 6.6$

$-0.05B = -0.6$

$B = -0.6/-0.05 = 12$

$B = 12$

Then substituting for A

$A = 22 - 12 = 10$

She bought 12 Brand B, and 10 candies Brand A

Age Problems

> Stop at every comma or period and ask yourself:
>
> "What is going on?" and then translate.

Let's do an example:

A) Janie is 10 years younger than her sister. If the sum of their ages is 50, how old is each sister?

Let's do it by translating the problem:

Janie is 10 years younger than her sister.

J = S − 10

If the sum of their ages is 50,

Sister + Janie = 50

Now you can put both equations together.

$J = S - 10$
$S + J = 50$

Let's substitute J into the second equation

$$J = S - 10$$
$$S + (S - 10) = 50$$

Solve for S:
2S - 10 = 50
2S = 50 + 10
S = 60/2 = 30

Her sister is **30 years old. Therefore, Janie is 20 years old.**
J = 30 - 10 = 20

> When age problems involve the present, past or the future, it is helpful to build a chart.

When you have age problems that involve the present, past or the future, it is helpful to build a chart.

Let's do an example:

A) Matt is twice as old as Luly. Three years from now, the sum of their ages will be 60. How old are Matt and Luly?

Build a chart this way:

- Ask yourself: how old is Matt today? how old is Luly today?

| TODAY | Luly's age: L | Matt: 2L |

Let's finish the chart.
In 3 years, how old is Luly? How old is Matt?

| TODAY | Luly's age: L | Matt: 2L |

| IN 3 YEARS | Luly in 3 years: L + 3 | Matt in 3 years: 2L + 3 |

In 3 years, the sum of their ages will be 60

Luly + Matt = 60

(L+3) + (2L + 3) = 60

3L + 6 = 60
3L = 54
L = 54/3 = 18

Luly is 18 and Matt is 2L = 18(2) = 36

In three years, Luly = 18 + 3 = 21, Matt = 2(18) + 3 = 39,
if you add 21 + 39 = 60

Practice Word Problems Set 1

1) A yellow and a blue car leave from the same place and travel opposite directions. The yellow car is traveling 50 mph while the blue car is traveling 70 mph. In how many hours will they be 1080 miles apart?
   a) 9 hours
   b) 10 hours
   c) 11 hours
   d) 12 hours

2) A fast car is traveling at an average rate of 60 miles per hour from town A to B in 5 hours. If it had traveled at 30 miles per hour, how **many more minutes** would it have taken to make the trip?
   a) 600 minutes
   b) 300 minutes
   c) 450 minutes
   d) 240 minutes

3) A car drives 40 mph from point A to B, and then returns from point B to A driving 60 mph. If the route AB= 100 miles. What is the average rate for the entire round trip?
   a) 50 mph
   b) 100 mph
   c) 48 mph
   d) 60 mph

4) How many liters of a 65% acid solution must be mixed with 45 liters of a 15% solution to obtain a solution that is 35% acid?
   a) 50
   b) 20
   c) 45
   d) 30

5) Amy, Bob, and Camila have 36 marbles. Amy has twice as many marbles as Bob, while Camila has three times as many marbles as Amy.
   How many marbles does Camila have?
   a) 4
   b) 8
   c) It cannot be determined with the information given
   d) 24

6) Emma has twice as many marbles as Roby, together they have 21 marbles. If she sold all her marbles to Roby for $18.20, What is the price of Emma's marbles?
   a) $11.20
   b) $9.20
   c) $2.60
   d) $1.30

7) The product of two numbers is 60. The first number is greater than the second number. If the average (arithmetic mean) of both numbers is twice the difference of such numbers. Which one could be the value of one of the numbers?

a) 6
b) 36
c) 13
d) 15

8) It takes Nana and Nina 4 hours to clean a house together. If Nana works alone, she can clean the house in "H" hours, while Nina can clean the house two times faster than Nana working alone. How long it will take to Nina to clean the house alone?

a) 6 hours
b) 24 hours
c) 12 hours
d) 10 hours

9) How many pounds of candy worth $3.20 a pound must be mixed with 5 pounds of candy worth $1.20 a pound to produce a mixture worth $2.2 a pound?

a) 6
b) 9
c) 5
d) 15

10) A company sells 30 watches every hour. Each watch costs $2.50 a piece to manufacture. If the watch's price is 25% more than the cost, what is the profit of the company in 2 hours?

a) $37.5

b) $18.75

c) $32.50

d) $13.76

11) Due to the economy, a building has "n" vacant units. Since the total monthly fixed cost of maintaining the building is $320 per unit, the rest of the building owners are responsible for paying an extra fee of $960. If the building has a total of 15 units, how many units are vacant?

a) 3

b) 5

c) 10

d) 12

12) A phone company with terrible customer service charges a flat fee of $30 monthly for the first 117 minutes used, and $0.32 for each additional minute. If Lupe's bill for the month was $72.88. How many minutes did she used?

a) 129

b) 134

c) 251

d) 210

13) Kristine wants to buy an expensive purse for $408. She goes to the bank and puts $300 on a saving account that yields a 12% yearly rate.

How long does she have to leave the money in the bank in order to reach her goal?

a) 7 years

b) 5 years

c) 3 years

d) forever

14) Maria has twice as many coins as Juan. Juan has five more coins that Lili. If the three friends have an average (arithmetic mean) of 45 coins. How many coins does Maria have?

a) 30

b) 35

c) 70

d) it cannot be determined

15) Vivian is 10 years older than her sister Judith. If in 20 years the sum of their ages is 64. How old is Vivian today?

a) 7

b) 17

c) 27

d) 37

16) Nicole is 7 years older than Adam. If in 5 years she will be twice as old as Adam, how old is Nicole today?

a) 2
b) 7
c) 9
d) 14

17) Oswald is three years younger than Virginia. If 11 years ago Virginia was twice as old as Oswald, what is the average mean (arithmetic mean) of their ages today?

a) 16
b) 15.5
c) 16.5
d) 17

18) If X is a real positive number such that $X^2 = 25$, then what is $X^3 - \sqrt{x}$?

a) $125\sqrt{5}$
b) $125 - \sqrt{5}$
c) 125
d) $5 - \sqrt{5}$

19) $3^{3x} + 3^{3x} + 3^{3x} =$

a) 3^{3x+1}

b) 3^{3x}

c) 9^{3x}

d) 27^{3x}

20) If an integer is selected from the integers 11-27 inclusive, what is the probability that the selected integer is a prime number?

a) 4/17

b) 5/16

c) 1/17

d) 5/17

21) A machine can print 40 pages in 5 minutes. At that constant rate, how many pages will 4 such machines print in 4 hours?

a) 1850

b) 1920

c) 6789

d) 7680

22) A secret recipe consists of sugar, butter, and flour mixed in the ratio 7:3:5, respectively, by weight. If the flour in the mixture weighs 25.2 pounds, how many pounds does **the total mixture** weigh?

    a) 18.85
    b) 54
    c) 74.6
    d) 75.6

23) A lock combination is made using 2 letters followed by three numbers. How many lock's combination can be made if repetition is NOT allowed?

    a) (27)(26)(10)(9)(9)
    b) (26)(25)(9)(7)(8)
    c) (26)(25)(10)(9)(8)
    d) (10)(10)(26)(26)(26)

24) Dianna has 6 designs each of which can be made with short or long sleeves. There are 5 color patterns available. How many different types of shirts can Dianna sell?

    a) 30
    b) 60
    c) 50
    d) 12

25) How many different 8-digit phone numbers are possible if the first digit cannot be 0 or 1?

   a) 80,000,000
   b) 10,000,000
   c) 90,000,000
   d) 120,000,000

26) How many different 4-digit sequences can be formed using the digits 1,2,3,4 if repetition of digits is allowed?

   a) 16
   b) 24
   c) 64
   d) 256

27) How many different 4-digit sequences can be formed using the digits 1,2,3,4 if repetition of digits is NOT allowed?

   a) 12
   b) 24
   c) 64
   d) 256

28)

$$\text{If } a = \frac{3 + \frac{1}{2}}{3 - \frac{4}{3}} \text{ and } b = \frac{2 - \frac{1}{5 - (-3 - 1)}}{-5}$$

What is the ratio of a to b?

a) -189/34

b) -114/34

c) 67/134

d) -1/67

29) The following data represents the amount of emails Cleo received during the week. What is the mean, median, and range of the data?

Days	Emails
Monday	2
Tuesday	7
Wednesday	4
Thursday	5
Friday	2
Saturday	0
Sunday	1

a) 3,2,7

b) 7,2,3

c) 3,1,2

d) 4,2,2

30) Simplify:

$$\left(\frac{4a}{Y^{-2}}\right)^2 \div \left(\frac{2a^{-1}}{Y^2}\right)^{-2}$$

a) 64

b) 64Y²

c) 16a⁴y²

d) 1

Answers to Set 1

1) A
2) B
3) C
4) D
5) D
6) D
7) A
8) A
9) C
10) A
11) A
12) C
13) C
14) C
15) B
16) C
17) B
18) B
19) A
20) D
21) D
22) D
23) C
24) B
25) A
26) D
27) B
28) A
29) A
30) A

Practice Test 1 (Basic Math)

Simplify

1. $-3 - 8 =$

   (A) −5   (B) −11   (C) 11   (D) 5

2. $-8 + 15 + 20 - 15 =$

   (A) 12   (B) −38   (C) −12   (D) 15

3. $-3(-2) =$

   (A) −5   (B) 2   (C) 1   (D) 6

4. $(-2)(-2)(-3) =$

   (A) 12   (B) 0   (C) −12   (D) −7

5. $|-2 + 9 - 15| =$
   (A) −8   (B) 8   (C) 9   (D) 26

6. $-|-10 + 2| =$
   (A) −5   (B) −12   (C) 12   (D) −8

7. $8 \times (4 + 1) - 7 \times 2 =$

   (A) 26   (B) 54   (C) −26   (D) 14

8. $-[(3+5) \times (-10+4) - 3] =$
   (A) −72 (B) −51 (C) 51 (D) −115

9. $(16 + 20) \div 3 \times 2 =$
   (A) −6 (B) 24 (C) 6 (D) 15

10. $3 - (4+2)^2 - (4 - (8 \div 2))^2 \times 2 =$
    (A) 1 (B) 0 (C) −33 (D) 23

11. $-3^2 - [8 - (4^2 - 10)] =$
    (A) 11 (B) −12 (C) 38 (D) −11

12. $-2^3 =$
    (A) −6 (B) 6 (C) −8 (D) 10

13. $\dfrac{9 - 1^2}{64} =$
    (A) $\dfrac{1}{10}$ (B) $-\dfrac{1}{8}$ (C) $\dfrac{1}{8}$ (D) $-\dfrac{9}{64}$

14. $(-2)^4 =$
    (A) −16 (B) 16 (C) 20 (D) −8

15. $\dfrac{2}{15} + \dfrac{1}{30} =$
    (A) $\dfrac{1}{6}$ (B) 6 (C) $-\dfrac{1}{6}$ (D) $\dfrac{2}{45}$

16. $\dfrac{2}{10} - \dfrac{9}{20} =$

(A) −5   (B) −10   (C) 5   (D) $-\frac{1}{4}$

17. $-\frac{10}{9} \times \frac{2}{4} =$

(A) $-\frac{5}{9}$ (B) $\frac{5}{9}$ (C) $-\frac{11}{18}$ (D) $-\frac{8}{13}$

18. $1\frac{1}{3} \div 2\frac{1}{4} =$

(A) 27   (B) $\frac{16}{27}$ (C) $-\frac{16}{27}$ (D) −1

19. $\frac{21}{5} \div \frac{7}{25} =$

(A) $\frac{147}{125}$ (B) 15   (C) 20   (D) −8

20. Find the LCM of 12 and 20.
(A) 60   (B) 100   (C) 15   (D) 13

21. Find the factors of 36
(A) 1,2,3,4,6,9,12,18,36   (B) 1,2,3,5,6   (C) 1,36   (D) 1,2,16,36

22. .48 × 9.3 =
(A) 4.464   (B) 0.46   (C) 4.9   (D) 4464

23. $-\frac{36}{54} =$

(A) $-\frac{2}{3}$ (B) $-\frac{3}{2}$ (C) −2   (D) $\frac{2}{3}$

24. $\frac{n}{5} = \frac{2}{3}$, Find value of n.

(A) $3\frac{1}{3}$ (B) $-\frac{10}{3}$ (C) -10 (D) 5

25. Find 36% *of* 90
    (A) 324   (B) 32.40   (C) 32.41   (D) 3240

26. $\frac{x}{4} = \frac{-3}{10}$, Find value of x.

    (A) 5 (B) $1\frac{1}{5}$ (C) $\frac{5}{6}$ (D) $-\frac{6}{5}$

27. What is 10% of 13 ?
    (A) 1.30   (B) 130%   (C) 23%   (D) .20

28. $\frac{x+3}{5} = \frac{3}{2}$, Find value of x .

    (A) $\frac{9}{2}$ (B) $-\frac{9}{2}$ (C) 10 (D) $\frac{2}{9}$

29. Convert $\frac{1}{4}$ to decimal.

    (A) 25   (B) 25%   (C) 2.5   (D) .25

30. $\frac{1}{3} + \frac{2}{9} + \frac{1}{27} =$

    (A) $\frac{16}{27}$ (B) $\frac{27}{16}$ (C) 27 (D) -16

Practice Test 1 answer key

1. B $-3 - 8 = -11$

2. A
   $-8 + 15 + 20 - 15 = 12$

3. D $-3(-2) = 6$

4. C $(-2)(-2)(-3) = -12$

5. B
   $|-2 + 9 - 15| = |-8| = 8$

6. D $-|-10 + 2| = -|-8| = -8$

7. A $8 \times (4 + 1) - 7 \times 2 = 8 \times 5 - 14 = 40 - 14 = 26$

8. C $-[(3 + 5) \times (-10 + 4) - 3] = -[8 \times (-6) - 3] = -[-48 - 3] = -[-51] = 51$

9. B $(16 + 20) \div 3 \times 2 = 36 \div 3 x 2 = 12 x 2 = 24$

10. C $3 - (4 + 2)^2 - (4 - (8 \div 2))^6 \times 2 = 3 - (6)^2 - (4 - (4))^6 \times 2 =$
    $3 - 36 - (0)^6 \times 2$
    $= 3 - 36 = -33$

11. D $-3^2 - [8 - (4^2 - 10)] = -3^2 - [8 - (6)] = -9 - [8 - 6]$
    $-9 - [2] = -11$

12.C $-2^3 = -2 \times -2 \times -2 = \mathbf{-8}$

13.C $\frac{9-1^2}{64} = \frac{9-1}{64} = \frac{8}{64} = \mathbf{\frac{1}{8}}$

14.B $(-2)^4 = (-2)(-2)(-2)(-2) = \mathbf{16}$

15.A $\frac{2}{15} + \frac{1}{30} = \frac{(2)2}{(2)15} + \frac{1}{30} = \frac{4}{30} + \frac{1}{30} = \frac{5}{30} = \mathbf{\frac{1}{6}}$

16. D.

$\frac{2}{10} - \frac{9}{20} = \frac{(2)2}{(2)10} - \frac{9}{20} = \frac{4}{20} - \frac{9}{20} = -\frac{5}{20} = \mathbf{-\frac{1}{4}}$

17.A $-\frac{10}{9} \times \frac{2}{4} = \frac{-20}{36} = \mathbf{\frac{-5}{9}}$

18.B $1\frac{1}{3} \div 2\frac{1}{4} = \frac{4}{3} \div \frac{9}{4} = \frac{4}{3} + \frac{4}{9} = \mathbf{\frac{16}{27}}$

19.B $\frac{21}{5} \div \frac{7}{25} = \frac{21}{5} \times \frac{25}{7} = \frac{21}{1} \times \frac{5}{7} = \frac{105}{7} = \mathbf{15}$

20.A Find the LCM of 12 and 20.

Find the prime factorization of 12 and 20 then the Least Common multiple is the product of the common and non-common prime factors with the highest exponent.

$2^2 \times 3 = 12$
$2^2 \times 5 = 20$ LCM=$2^2 \times 3 \times 5 = \mathbf{60}$

21.A The factors of 36.

The factors of 36 are: **1,2,3,4,6,9,12,18,36**

22.A $0.48 \times 9.3 = \mathbf{4.464}$

23.A $\frac{-36}{54} = \frac{-2}{3}$

24.A $\frac{n}{5} = \frac{2}{3}$ $\frac{n}{5} = \frac{2}{3}$ $3n = 10$

$\frac{3n}{3} = \frac{10}{3}$ $n = \frac{10}{3} = 3\frac{1}{3}$

25.B $36\% \; of \; 90$ $0.36 \times 90 = = \mathbf{32.40}$

26.D $\frac{x}{4} = \frac{-3}{10}$ Cross multiplying $10x = -12$

$\frac{10x}{10} = \frac{-12}{10} = -6/5$ $x = \frac{-6}{5}$

27.A   What is 10% of 13. $0.10 \times 13 = \mathbf{1.30}$

28.A $\frac{x+3}{5} = \frac{3}{2}$ Cross multiplying $2(x+3) = 15$

$2x + 6 = 15$ $2x = 15 - 6$

$2x = 9$ $\frac{2x}{2} = \frac{9}{2}$ $x = \frac{9}{2}$

29.D   Convert $\frac{1}{4}$ to decimal. ¼ = **0.25**

30.A $\quad \frac{1}{3}+\frac{2}{9}+\frac{1}{27} = \frac{(9)1}{(9)3}+\frac{(3)2}{(3)9}+\frac{1}{27} = \frac{9}{27}+\frac{6}{27}+\frac{1}{27} = \frac{9+6+1}{27} = \frac{16}{27}$

Practice Test 2

1. $-4A + 8B$ if $A = -3,\quad B = -1$
   (A) 3   (B) -3 (C) 4   (D) -4

2. $-2X - 5Y + Y$ if $Y = 3,\quad X = 0$
   (A) -12 (B) 12   (C) -10 (D) 25

3. $3P^{23} + 4T^{33}$ if $T = 1,\quad P = -1$
   (A) 1   (B) 7   (C) -1 (D) -7

4. $\frac{RRR}{EE}$ if $R = -2,\quad E = 8$
   (A) $\frac{1}{8}$ (B) 8   (C) $\frac{-1}{8}$ (D) 64

5. $Y + Y + Y + Y$
   (A) Y^4 (B) $4Y$ (C) $1Y$ (D) $-4Y$

6. $5ab + 9b - 6ba + b$
   (A) $-ab + 10b$ (B) $10b + ab$ (C) $11ab + b$ (D) b

7. $-3X + 9Y - 10X$
   (A) $-13X + 9Y$ (B) $9Y + 13X$ (C) $13X$ (D) $12XY$

8. $9X + 6(X - 3) - 5X + 3$
   (A) $X - 15$ (B) $10X - 15$ (C) $-25X$ (D) -15

9. $(-(-7^0))$
   (A) -1 (B) 0 (C) -7 (D) 1

10. $\frac{1}{X^{-9}}$
    (A) X^{-9} (B) -9 (C) X^9 (D) $\frac{1}{X^9}$

11. $(Y^2 + 2)^2$
    (A) $Y^2 + 4Y^2 + 4$ (B) $Y^4 + 8Y^2$ (C) $Y^2 + 4$ (D) $Y^4 + 4Y^2 + 4$

12. $8Y^3(-Y^5 + 5P) - 2P(-1 + Y^8)$
    (A) $-8Y^8 + 40PY^3 + 2P - 2PY^8$ (B) $-8Y^8 + 40$
    (C) 0 (D) $2PY^3$

13. $\frac{6X-18}{2X-12}$
    (A) $X - 6$ (B) $\frac{2}{9}$ (C) $3\frac{x-3}{x-6}$ (D) 0

14. $\frac{X^{-4}}{X^{-8}}$
    (A) X^{-12} (B) X^4 (C) $\frac{1}{X^4}$ (D) $\frac{1}{X^{-12}}$

15. $-3 - |-2 + 5|$
    (A) -6 (B) -9 (C) -11 (D) 6

16. $\frac{2X^3}{-16X^5}$

(A) $8X^2$ (B) $-8X^2$ (C) $\frac{1}{-8X^2}$ (D) $-18X^2$

17. $\sqrt{27x^{11}y^{17}}$
    (A) $3x^5y^8\sqrt{3xy}$ (B) $xy\sqrt{3xy}$ (C) $9x^5y^8\sqrt{3xy}$ (D) $27xy$

18. $3\sqrt{7x^{19}} + 4x^9\sqrt{7x}$
    (A) $7x^9\sqrt{7}$ (B) $7x\sqrt{7}$ (C) $12x^9\sqrt{7}$ (D) $7x^9\sqrt{7x}$

19. $-(3X^2Y^3)^2$
    (A) $-6x^4y^6$ (B) $-9X^4Y^6$ (C) $-9X^4Y^5$ (D) $9X^4Y^6$

Solve for the indicated Variable

20. $AXY = J$ for "A"
    (A) $A = JXY$ (B) $A = \frac{XY}{J}$ (C) $A = \frac{J}{XY}$ (D) $A = 1$

21. $\frac{J}{K} = L$ for "K"
    (A) $K = \frac{J}{L}$ (B) $K = \frac{L}{J}$ (C) $K = LJ$ (D) $JK = L$

22. $PX + CX = Y$ for "X"
    (A) $X = \frac{Y}{P+C}$ (B) $X = P + C$
    (C) $X = \frac{P+C}{Y}$ (D) $X = P + C + Y$

Translate the following expressions:

23.   5 less than X is twice the sum of X and 3.
    (A) $X - 5 = 2(X + 3)$ (B) $5 - X = 2(X + 3)$
    (C) $5 - X = 2(X - 3)$ (D) $X - 5 = 2$

24. Vanessa is 18 years older than Peter
    (A) $V + 18$ (B) $V = 18 + P$ (C) $V + P = 18$ (D) $VP = 18$

25. $(6Z^{-5}X^2)^{-2}$

    (A) $\dfrac{1}{36Z^5X^4}$ (B) $-36Z^{10}X^4$ (C) $\dfrac{Z^{10}}{36X^4}$ (D) $36Z^{10}X^{-4}$

26. $\dfrac{-64X^{-4}Y}{32XY}$

    (A) $\dfrac{-8Y}{4X^5}$ (B) $\dfrac{1}{-2X^5}$ (C) $\dfrac{-2}{X^5}$ (D) $\dfrac{X^5}{-2}$

27. $\left(\dfrac{X^{-3}Y^8}{X^6Y^5}\right)^{-2}$

    (A) $\dfrac{X^{18}}{Y^6}$ (B) $\dfrac{Y^6}{X^{18}}$ (C) $X^{18}Y^6$ (D) $-X^{18}Y^6$

28. $-\left(\dfrac{J^{-9}K^2}{K^8J^{-3}}\right)^0$

    (A) -1 (B) 0 (C) 1 (D) $J^{-17}K^{10}$

29. $\sqrt{2}(\sqrt{2} + 3)$
    (A) $-5\sqrt{2}$ (B) $5\sqrt{2}$ (C) $4 + 3\sqrt{2}$ (D) $2 + 3\sqrt{2}$

30. If the length of a sign is 10 inches longer than its width. Find the width if the perimeter is 44 inches.

(A) $w = 6\ inches$ (B) $w = 16\ inches$

(C) $w = 3\ inches$ (D) $w = 7.5\ inches$

Practice Test 2 answer key

1. C $-4A + 8B$ if $A = -3,$ $B = -1$
$-4(-3) + 8(-1) = \mathbf{4}$

2. A $-2X - 5Y + Y$ if $Y = 3,$ $X = 0$
$-2(0) - 5(3) + (3) = \mathbf{-12}$

3. A $3P^{23} + 4T^{33}$ if $T = 1,$ $P = -1$
$3(-1)^{23} + 4(1)^{33} = \mathbf{1}$

4. C $\dfrac{RRR}{EE}$ if $R = -2,$ $E = 8$

$\dfrac{(-2)(-2)(-2)}{(8)(8)} = \dfrac{\mathbf{-1}}{\mathbf{8}}$

5. B $Y + Y + Y + Y = \mathbf{4Y}$

6. A $5ab + 9b - 6ba + b = \mathbf{-ab + 10b}$

7. A $-3X + 9Y - 10X = \mathbf{-13X + 9Y}$

8. B $9X + 6(X - 3) - 5X + 3 = \mathbf{10X - 15}$

9. A $-(-7^0) = -(-1) = \mathbf{1}$

10. C $\dfrac{1}{X^{-9}} = \mathbf{X^9}$

11. D $(Y^2 + 2)^2 = (Y^2 + 2)(Y^2 + 2) = Y^4 + 2Y^2 + 2Y^2 + 4$
$= \mathbf{Y^4 + 4Y^2 + 4}$

12.A $\quad 8Y^3(-Y^5+5P)-2P(-1+Y^8) = \mathbf{-8Y^8+40PY^3+2P-2PY^8}$

13.C $\quad \dfrac{6x-18}{2x-12} = \dfrac{6(X-3)}{2(X-6)} = \mathbf{3\dfrac{x-3}{x-6}}$

14.B $\quad \dfrac{X^{-4}}{X^{-8}} = \dfrac{X^8}{X^4} = \mathbf{X^4}$

15.A $\quad -3-|-2+5| = -3-|3| = -3(-3) = \mathbf{-6}$

16.C $\quad \dfrac{2X^3}{-16X^5} = \dfrac{\cancel{2}\cancel{X}.\cancel{X}.X.}{-1\cancel{6}\cancel{X}.\cancel{X}.X.X.X} = \dfrac{1}{\mathbf{-8X^2}}$

17.A $\quad \sqrt{27x^{11}y^{17}} = \sqrt{9}\sqrt{3x^{10}x^1 y^{16}y^1} = \mathbf{3x^5 y^8 \sqrt{3xy}}$

18.D $\quad 3\sqrt{7x^{19}} + 4x^9\sqrt{7x} = 3\sqrt{7x^{18}x} + 4x^9\sqrt{7x} = 3x^9\sqrt{7x} + 4x^9\sqrt{7x} = \mathbf{7x^9\sqrt{7x}}$

19.B $\quad -(3X^2 Y^3)^2 = -(3^2 X^{2x2} Y^{3x2}) = \mathbf{-9X^4 Y^6}$

20.C $\quad AXY = J \qquad$ for "A"

$\dfrac{A\cancel{XY}}{\cancel{XY}} = \dfrac{J}{XY} = \mathbf{A = \dfrac{J}{XY}}$

21.A $\quad \dfrac{J}{K} = L \qquad$ for "K"

$\dfrac{(K)J}{K} = \dfrac{L(K)}{1} \qquad \dfrac{J}{L} = \dfrac{K\cancel{L}}{\cancel{L}} = \dfrac{J}{L} = K$

$\mathbf{k = \dfrac{J}{L}}$

22.A $\quad PX + CX = Y$ for "X"

Factor the letter X(P+C)= Y

$$\frac{X(P+C)}{P+C} = \frac{Y}{P+C} \qquad X = \frac{Y}{P+C}$$

23.A 5 less than X is twice the sum of X and 3.
$$X - 5 = 2(X + 3)$$

24.B Vanessa is 18 years older than Peter
$$V = 18 + P$$

25.C $(6Z^{-5}X^2)^{-2} = (6^{-2}Z^{10}X^{-4}) = \dfrac{Z^{10}}{36X^4}$

26.C $\dfrac{-64X^{-4}Y}{32XY} = \dfrac{-2Y}{XX^4Y} = \dfrac{-2}{X^5}$

27.A $\left(\dfrac{X^{-3}Y^8}{X^6Y^5}\right)^{-2} = \left(\dfrac{X^6Y^{-16}}{X^{-12}Y^{-10}}\right) = \left(\dfrac{X^6X^{12}Y^{10}}{Y^{16}}\right) = \left(\dfrac{X^{18}}{Y^6}\right)$

28.C $-\left(\dfrac{J^{-9}K^2}{K^8J^{-3}}\right)^0 = -1$. Remember the (-) is outside

29.D $\sqrt{2}(\sqrt{2} + 3) = \sqrt{4} + 3\sqrt{2} = 2 + 3\sqrt{2}$

30.A If the length of a sign is 10 inches longer than its width. Find the width if the perimeter is 44 inches

The lenght is $L = 10 + W$.   **The perimeter formula**
P=2L+2W. Replacing L into the formula
44=2((10+W) + 2W
44=20+2W+2W
24=4W
W=24/4= **6 inches**

Practice Test 3 (Intermediate Algebra)

1. $2|2X+5| < 12$

    (A) $(0, -6)$ (B) $\left(-\frac{11}{2}, \frac{1}{2}\right)$ (C) $[-6, 0]$ (D) $[-6, 0]$

2. $|3X + 8| = -3$

    (A) -11 (B) No Solution   (C) 11 (D) $\frac{11}{3}$

3. $-X \geq -4$

    (A) $(4, 0)$ (B) $[-4, \infty)$ (C) $(4, \infty)$ (D) $(-\infty, 4]$

4. $3(X + 2) \geq -18$

    (A) $[-8, \infty)$ (B) $(4, 0)$ (C) $(\infty, -4]$ (D) $[\infty, -4]$

5. $9 < -3X < 27$

    (A) $-1 > X > -9$ (B) $-3 < X < 9$ (C) 1 (D) $-9 < X < -3$

6. $6 < \frac{2}{3}X \leq 8$

    (A) $18 \leq X \leq 24$ (B) $9 \leq X \leq 12$ (C) $\frac{2}{3}X$

    (D) $9 < X \leq 12$

7. $P = \frac{1}{3}MV^2$ for "V"

    (A) $V = \sqrt{\frac{3P}{M}}$ (B) $V = \sqrt{\frac{M}{3P}}$ (C) $V^2 = 3P$ (D) $\sqrt{V+P}$

Find the Slope

8. $(6,0)(6,4)$

 (A) Undefined (B) $\frac{1}{4}$ (C) 0 (D) 1

9. $(-4,3)(-8,5)$

 (A) $\frac{2}{1}$ (B) $\frac{-1}{2}$ (C) $\frac{1}{2}$ (D) 4

10. $3X - 4Y = -24$ Find the X and Y intercepts.

 (A) $(-8,0), (6,0)$ (B) $(8,0), (6,0)$
 (C) $(0,6), (-8,0)$ (D) $(12,0), (0,-3)$

11. Build the equation of the line in point-slope form with slope $m = 5$ passing through $(3,2)$

 (A) $3 - Y = 5(2 - X)$ (B) $Y - 2 = 5(X - 3)$
 (C) $Y = 5$ (D) $Y - 2 = 5$

12. Convert the line into slope-intercept form $\frac{2}{3}Y + 3X = 5$

 (A) $Y = -9X + 15$ (B) $Y = 9X + 15$ (C) $Y = \frac{-9}{2}X + \frac{15}{2}$ (D) $Y = -3X$

13. Convert into Standard Form. $Y = 4X + 5$

 (A) $4X + Y = 5$ (B) $-4X - Y = 5$ (C) $4X - Y = -5$ (D) $4X - Y = 5$

14. Convert into Slope-Intercept form: $Y - 3 = \frac{1}{2}(X + 3)$

 (A) $Y = \frac{1}{2}X - 9/2$ (B) $Y = \frac{1}{2}X + 9/2$

(C) $Y = -\frac{1}{2}X + 9/2$ (D) $Y = \frac{1}{3}X + 9/2$

15. Solve for "X". $\frac{X+3}{-4} = \frac{1}{2} + x$

(A) -1 (B) 1   (C) 5   (D) -10

16. $\frac{\frac{1}{4}}{\frac{3}{7}}$

(A) $\frac{7}{12}$ (B) $\frac{-12}{7}$ (C) $-\frac{7}{12}$ (D) 6

17. $\frac{\frac{2}{X}+\frac{3}{Y}}{\frac{4}{XY}+\frac{5}{YX}}$

(A) $9 + 2Y$ (B) $\frac{9}{2Y+3X}$ (C) $\frac{2Y+3X}{9}$

(D) $\frac{2Y+3X}{9xy}$

18. $\frac{\frac{-3}{X}+\frac{2}{Y}}{4+\frac{3}{XY}}$

(A) $4XY + 3$ (B) $\frac{-3Y+2X}{4XY+3}$

(C) $\frac{4XY+3}{-3Y+2X}$ (D) $-3Y + 2X$

19. Find the line that is perpendicular to $Y = -\frac{1}{3}X + 1$ that passes through (-2,4)

(A) $Y = \frac{-1}{3}X + 10$ (B) $Y = -3X + 1$

(C) $Y = 3x - 10$ (D) $Y = 3X + 10$

20. Build the equation of the line if the slope is $m = 2$ and passes through (3,1)

(A) $Y = -2X + 6$ (B) $Y = 2X - 5$ (C) $Y = \frac{2}{3}X - 4$ (D) $Y = 2X - 8$

21. Solve for "Y" $-3x - \frac{2}{4}y = 6$

(A) $Y = -6X - 12$ (B) $Y = 6X + 12$ (C) $Y = 12 + \frac{2}{3}X$ (D) $3X + \frac{2}{4}Y = 6$

22. Solve for X: $\frac{3}{4}X + \frac{1}{16}X = -1$

(A) $X = 13$ (B) $X = -\frac{13}{16}$ (C) $X = \frac{-16}{13}$ (D) $X = \frac{-16}{15}$

23. Graph $Y = \frac{3}{2}X - 6$

(A) 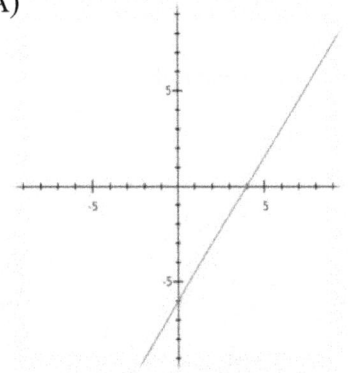   (B)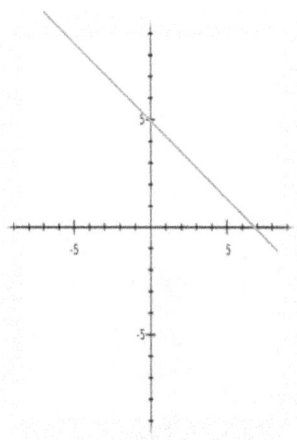

(C) 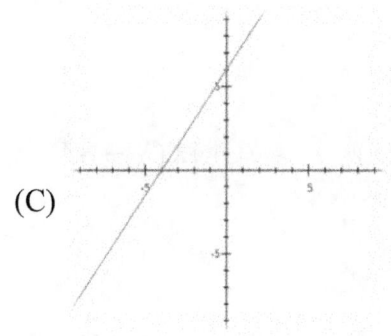 (D)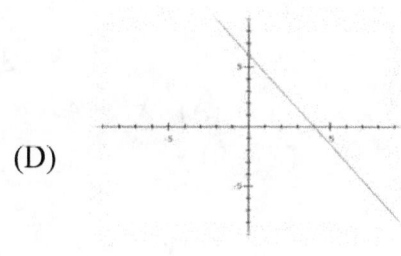

24. Solve for X: $-3 = \frac{1}{X+3}$

(A) $X = 10$ (B) $X = \frac{3}{10}$ (C) $X = -\frac{10}{3}$ (D) $X = 3$

25. Solve for X: $\frac{2}{3}X + 1 = \frac{4}{7}X$

(A) $X = 0$ (B) $X = 8\frac{1}{2}$ (C) $X = \frac{2}{21}$ (D) $X = \frac{-21}{2}$

26. Solve for X: $\frac{X-1}{3} + 2X = -1$

(A) $X = 5$ (B) $X = \frac{-2}{7}$ (C) $X = \frac{5}{2}$ (D) $X = 7$

27. $(X - 2)^3 =$
(A) $X^3 - 2$ (B) $X^3 - 2X^2 - 4X + 8$
(C) $X^3 - 6X^2 + 12X - 8$ (D) $X^2 - 4X + 8$

28. $-5X < -10$
(A) $X > 2$ (B) $X < -5$ (C) $X < 2$ (D) $X > 5$

Factor

29. $4X^2 - 32XY$

(A) $4X^2 - 4$ (B) $4X^2(8Y)$ (C) $4X(X - 8Y)$ (D) $(X - 8Y)$

30. $3X^2 - 27$

(A) $3(X + 3)(X - 3)$ (B) $3(X^2 - 27)$ (C) $X^2 - 30$ (D) $X^2 - 27$

Practice Test 3 answer key

1. B $2|2X+5| < 12$ $\frac{2}{2}|2X+5| < \frac{12}{2}$

   $2X+5 < 6$ and $-2X-5 < 6$

   $2X < 6-5$ and $-2X < 11$

   $\frac{2X}{2} < \frac{1}{2}$ $-\frac{2X}{-2} > 11/-2$

   $X < \frac{1}{2}$ and $X > \frac{-11}{2}$ (-11/2, 1/2)

2. B

   $|3X+8| = -3$

   (Absolute value equations can never have a negative solution)

3. D

   $-X \geq -4$ $X \leq 4$ $(-\infty, 4]$

4. A

   $3(X+2) \geq -18$ $3X+6 \geq -18$ $3X \geq -18 - 6$

   $3X \geq -24$ $\frac{3}{3}X \geq \frac{-24}{3}$ $X \geq -8$ **[-8, ∞)**

5. D $9 < -3X < 27$ $\left| \frac{9}{-3} < \frac{-3}{-3}X < \frac{27}{-3} \right.$ $-3 > X > -9$

   or

   $\boldsymbol{-9 < X < -3}$

6. D $6 < \frac{2}{3}X \leq 8$ $(3)6 < \left| \frac{(3)2}{3}X \leq 8(3) \right. \frac{18}{2} < \frac{2}{2}X \leq \frac{24}{2}$

$$9 < X \leq 12$$

7. A $P = \frac{1}{3}MV^2$ $(3)P = (3)\frac{1}{3}MV^2$ $\frac{3P}{M} = \frac{MV^2}{M}$

$$\sqrt{\frac{3P}{M}} = \sqrt{V^2}$$

$$V = \sqrt{\frac{3P}{M}}$$

8. A $(6,0)(6,4)$ $m = \frac{Y_2 - Y_1}{X_2 - X_1}$ $m = \frac{4-0}{6-6} = \frac{4}{0} = \mathbf{Undefined}$

9. B $(-4,3)(-8,5)$ $m = \frac{Y_2 - Y_1}{X_2 - X_1}$ $m = \frac{5-3}{-8-(-4)} = \frac{2}{-4} = \frac{-1}{2}$

10. C $3X - 4Y = -24$

    Y −INTERCEPT $3(0) - 4Y = -24$ $\frac{-4Y}{-4} = \frac{-24}{-4}$

    $Y = 6$ **(0,6)**

    X −INTERCEPT $3X - 4(0) = -24$ $\frac{3X}{3} = \frac{-24}{3}$

    $X = -8$ **(-8,0)**

11. B

    Using the formula $y - y1 = m(x2 - x1)$ $Y - 2 = 5(X - 3)$

12. C

    $\frac{2}{3}Y + 3X = 5$ Multiply every term by 3 $\frac{(3)2}{3}Y + (3)3X = (3)5$

    $\mathbf{2Y + 9X = 15}$

$$2Y = -9X + 15 \qquad \frac{2Y}{2} = \frac{-9X}{2} + \frac{15}{2} \qquad Y = \frac{-9}{2}X + \frac{15}{2}$$

13. C $Y = 4X + 5$ the standard form of the line is ax+by=c
$$-4X + Y = 5$$
$$-(-1)4X + (-1)Y = (-1)5$$
$$\mathbf{4X - Y = -5}$$

14. B $Y - 3 = \frac{1}{2}(X + 3)$, the slope intercept form is y=mx+b

   Multiply by 2 $2Y - 6 = 2 * \frac{1}{2}(X + 3)$ $2Y - 6 = X + 3$

   $$2Y = X + 6 + 3 \qquad 2Y = X + 9 \qquad Y = \frac{X}{2} + \frac{9}{2}$$

15. A

   $\frac{X+3}{-4} = \frac{1}{2} + x$ Multiply by -4 every term $-4\left(\frac{X+3}{-4}\right) = -4\left(\frac{1}{2}\right) - 4x +$

   $$(X + 3) = -2 - 4x$$
   $$x + 4x = -2 - 3 \qquad\qquad x + 4x = -2 - 3$$
   $$5x = -5 \quad x = -1$$

16. A

   $$\frac{\frac{1}{4}}{\frac{3}{7}} = \frac{1}{4} \div \frac{3}{7} = \frac{1}{4} \times \frac{7}{3} = \frac{7}{12}$$

17. C $\dfrac{\frac{2}{X}+\frac{3}{Y}}{\frac{4}{XY}+\frac{5}{YX}} = \dfrac{\frac{(Y)2}{(Y)X}+\frac{3(X)}{Y(X)}}{\frac{4}{XY}+\frac{5}{YX}} = \dfrac{\frac{2Y}{XY}+\frac{3X}{XY}}{\frac{4}{XY}+\frac{5}{YX}} = \dfrac{\frac{2Y+3X}{XY}}{\frac{9}{XY}} = \dfrac{XY(2Y+3X)}{XY(9)} = \dfrac{2X+3Y}{9}$

18. B $\dfrac{\frac{-3}{X}+\frac{2}{Y}}{4+\frac{3}{XY}} = \dfrac{\frac{-(Y)3}{(Y)X}+\frac{2(X)}{Y(X)}}{\frac{4(XY)}{1(XY)}+\frac{3}{XY}} = \dfrac{\frac{-3Y}{XY}+\frac{2X}{XY}}{\frac{4XY}{XY}+\frac{3}{XY}} = \dfrac{\frac{-3Y+2X}{XY}}{\frac{4XY+3}{XY}}$

$\dfrac{XY(-3Y+2X)}{XY(4XY+3)} = \dfrac{-3Y+2X}{4XY+3}$

19. D $Y = -\dfrac{1}{3}X + 1$

   The slope is m=-1/3 , then the perpendicular slope is
   m= 3. Using the point (-2,4)
   $y - 4 = 3(x + 2)$
   $y - 4 = 3x + 6$
   $y = 3x + 6 + 4$
   $y = 3x + 10$

20. B
   $m = 2$ (3,1)   Use $Y = mX + b$ $1 = 2(3) + b$
   $1 = 6 + b$ $1 - 6 = b$ $b = -5$
   Substituting into y=mx+b   **$Y = 2X - 5$**

21. A Convert $-3X - \dfrac{2}{4}Y = 6$ to $y = mx + 6$

   $-\dfrac{2}{4}Y = 3X + 6$ $-\dfrac{(4)2}{4}Y = (4)3X + 6(4)$ $-\dfrac{2Y}{2} = \dfrac{12X}{-2} + \dfrac{24}{-2}$

   $Y = -6X - 12$

22. C

$$\tfrac{3}{4}X + \tfrac{1}{16}X = -1 \qquad \tfrac{(16)3}{4}X + \tfrac{(16)1}{16}X = -1(16) \qquad 12X + X = -16$$

$$\tfrac{13X}{13} = \tfrac{-16}{13} \qquad X = \tfrac{-16}{13}$$

23. A $Y = \tfrac{3}{2}X - 6$, using the y-intercept (0,-6) and the slope m=3/2 (rise 3 and run 3) the answer is **graph A**

24. C $-3 = \tfrac{1}{X+3} \qquad -3(X+3) = \tfrac{1(X+3)}{X+3} \qquad -3X - 9 = 1$

-3x=1+9 $\qquad \tfrac{-3X}{-3} = \tfrac{10}{-3} \qquad X = -\tfrac{10}{3}$

25. D $\tfrac{2}{3}X + 1 = \tfrac{4}{7}X \qquad \tfrac{(21)2}{3}X - \tfrac{(21)4}{7}X = -1(21) \qquad 14X - 12X = -21$

$2X = -21 \qquad \tfrac{2X}{2} = \tfrac{-21}{2} \qquad X = \tfrac{-21}{2}$

26. B $\tfrac{X-1}{3} + 2X = -1 \qquad \tfrac{(3)(X-1)}{3} + (3)2X = -1(3) \qquad X - 1 + 6X = -3$

$\tfrac{7X}{7} = \tfrac{-2}{7} \qquad X = \tfrac{-2}{7}$

27. C
$(X-2)^3 = (X-2)(X-2)(X-2) \qquad =(X-2)(X^2 - 2X - 2X + 4)$
$= (X-2)(X^2 - 4X + 4) \qquad =X^3 - 4X^2 + 4X - 2X^2 + 8X - 8$
$=X^3 - 6X^2 + 12X$

28. A

$-5X < 10 \qquad \tfrac{-5}{-5}X < \tfrac{-10}{-5} \qquad X > 2$

Remember, when you divide by a negative number the signs flips.

29. C
$$4X^2 - 32XY = 4X(X-8Y)$$
30. A $\quad 3X^2 - 27 = 3(X^2-9) = 3(X+3)(X-3)$

Practice Test 4

1. Solve for X
$$\sqrt{3X-5} = 1$$
(A) No Solution (B) $X = 0$ (C) $X = 1$ (D) $X = 2$

2. Solve for the values of X
$$2X^3 + 5X^2 - 8X - 20 = 0$$
(A) $\{-\frac{5}{2}, -2, 2,\}$ (B) $\{-4, 4, \frac{5}{2}\}$ (C) $\{-1, 0, 2\}$

(D) $\{-2, 2, \frac{5}{2}\}$

3. If $F(X) = -3X^2 + X - 5$, Find $F(a-2)$
   (A) $\quad 3a^2 + 13a + 19$ (B) $-3a^2 + 13a - 19$
   (C) $-3a^2 - 5a - 5$ (D) None of the above.

4. Find the distance between $(-3, 5)$ and $(2, 6)$.
   Hint $(d = \sqrt{(Y_2 - Y_1)^2 + (X_2 - X_1)^2})$
   (A) $\sqrt{26}$ (B) $2\sqrt{3}$ (C) $\sqrt{29}$
   (D) $\sqrt{2}$

5. Find the domain of the function. $F(X) = \frac{3}{X+5}$
   (A) $(-\infty, -5) \cup (-5, \infty)$ (B) $(-\infty, -5] \cup [-5, \infty)$
   (C) $(-\infty, 5) \cup (5, \infty)$ (D) All real numbers.

6. Find the solution of the system of equations.

   $3X - Y = -2$

   $-X + 5Y = 24$

   (A) (5,1)   (B) (1,5)   (C) (−1,5)   (D) (−5,−1)

7. Solve. $-5 = 3X^2 - 8X$

   (A) $\left\{1, \frac{5}{3}\right\}$ (B) $\left\{-1, \frac{-5}{3}\right\}$ (C) $\left\{0, \frac{5}{3}\right\}$ (D) No Solution

8. Graph the following Inequalities.

   $Y \geq -3X + 5$

   $X + 2 \geq 0$

(A)   B)

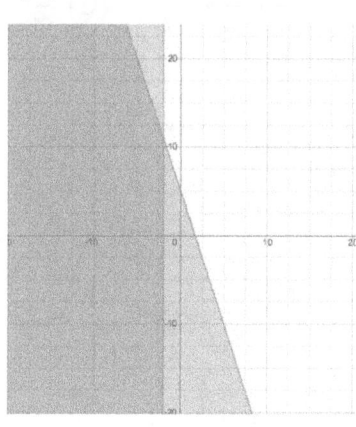   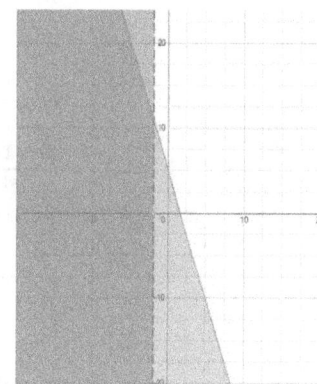

C)   D)

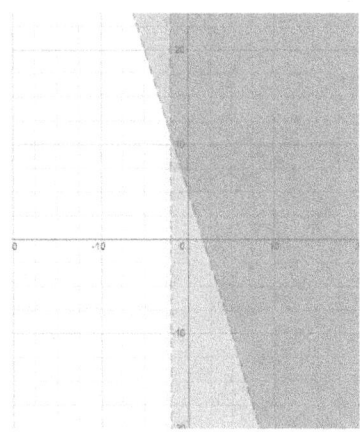
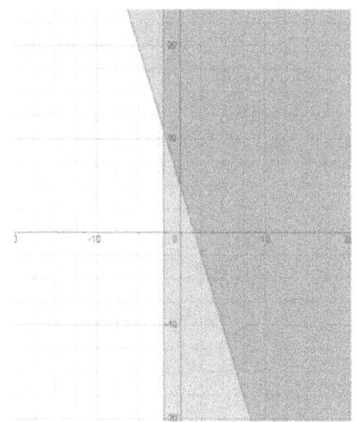

9. Simplify: $\dfrac{3}{X^2+3X-10} - \dfrac{X}{X^2+2X-15}$

(A) $\dfrac{X^2-5X+9}{(X+5)(X-2)(X-3)}$ (B) $\dfrac{-X^2+5X-9}{(X+5)(X-2)(X-3)}$ (C) $\dfrac{3X}{X+2}$

(D) $\dfrac{X}{X+3}$

10. Simplify: $\dfrac{8a+16}{4-a^2}$

(A) 8   (B) $\dfrac{8}{2+a}$ (C) $\dfrac{8}{2-a}$ (D) $\dfrac{8}{a}$

11. Find the X-intercepts of the function $Y = 2X^2 + 3X - 5$ (Hint: Make $Y = 0$)

(A) $(0,-5)(0,6)$ (B) $(0,1)(0,\dfrac{-5}{2})$ (C) $(-1,0)(\dfrac{5}{2},0)$

(D) $(1,0)(\dfrac{-5}{2},0)$

12. If $F(X) = 3X - 2$ and $G(X) = -5X + 6$, Find $F(X) - G(X)$.

(A) $2X + 8$ (B) $8X - 2$ (C) $8X - 8$ (D) $-8X + 4$

13. Graph the following line using the intersects: $2Y - 3X = 36$

(A) 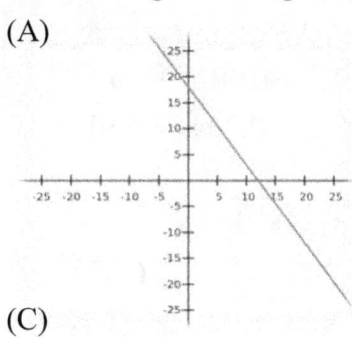 (B)

(C)

(D) None of the above

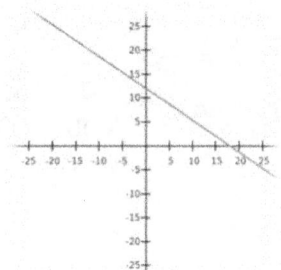

14. The equation of Kinetic Energy is $K_E = \frac{1}{2}MV^2$ (Kinetic Energy).

Solve for V.

(A) $V = \frac{M}{2K_E}$ (B) $V = \left(\frac{2M}{K_E}\right)^2$ (C) $V = \sqrt{\frac{M}{2K_E}}$ (D) $V = \sqrt{\frac{2K_E}{M}}$

15. If 3 is added to twice a number, the result is 5 less than 3 times the number. Find the resulting equation.

(A) $3 + 2X = 3X - 5$ (B) $-3 - 2X = 3X + 5$

(C) $3 + 2X = 5 - 3X$ (D) $2(3 + X) = 3X - 5$

16. Simplify $\quad \frac{a+b}{a-b} \div \frac{-a-b}{2a-2b}$

(A) −2   (B) $-\frac{1}{2}$ (C) $2a$ (D) $\frac{2a}{b}$

17. Simplify $(8X - 3Y)^2$
   (A) $64X^2 - 48XY + 9Y^2$ (B) $8X^2 - 9Y^2$
   (C) $64X^2 + 48XY + 9Y^2$ (D) $64X^2 + 9Y^2$

18. Simplify $(5X^2 - 2X + 3)(X - 1)$
   (A) $5X^3 - 7X^2 + 5X - 3$ (B) $5X^3 + 7X^2 - 5X + 3$
   (C) $5X^3 + 7X^2 - 5X - 3$ (D) $6X^3 - 7X^2 - 5X + 3$

19. Find the parallel slope of the line $X = -3$
   (A) $M_\parallel = -3$ (B) $M_\parallel = \frac{1}{3}$ (C) 0   (D) Undefined

20. $\frac{X^4 - 256}{X^2 - 4X - 32}$

   (A) $\frac{(X-4)(X+8)}{(X-8)}$ (B) $\frac{(X+4)(X^2+16)}{X-8}$ (C) $\frac{X-4}{X-8}$

   (D) $\frac{(X-4)(X^2+16)}{(X-8)}$

21. Evaluate the following expression
   $X^n \cdot X^m \cdot X^{-y}$ If $X = -1$, $n = 2$, $m = 3$, and $y = -5$
   (A) 1   (B) −1   (C) 0   (D) −3

22. Simplify: $\sqrt{729a^{17}b^{16}c^{11}}$
   (A) $27a^3b^2\sqrt{c}$ (B) $81\sqrt{abc}$
   (C) $27a^4b^4c\sqrt{ac}$ (D) $27a^8b^8c^5\sqrt{ac}$

23. Solve for "d" in the following formula $V_F^2 = V_I^2 + 2ad$
   (A) $d = \frac{V_F^2 - V_I^2}{2a}$ (B) $d = \frac{V_F^2 + V_I^2}{2a}$

(C) $d = \sqrt{\frac{V_F - V_I}{2a}}$ (D) None of the above

24. Solve the following equation using the quadratic formula:
$$3X^2 + 6X - 1 = 0$$

(A) $X = \frac{-3 \pm 4\sqrt{3}}{3}$ (B) $X = \frac{-3 \pm 2\sqrt{3}}{3}$

(C) $X = \frac{\sqrt{3}}{3}$ (D) $X = \frac{3 \pm 4\sqrt{3}}{3}$

25. Find the number of solutions for the following quadratic equation :
$$X^2 + 3X + 4 = 0$$
(Hint: Use the discriminant)

(A) No Real Solutions (B) 1 Real Solution
(C) 2 Real Solutions (D) One real and one no real solution

26. If $F(X) = 3X^3 + 6X^2 - 1$, Find $F(-1)$
(A) -2 (B) 2 (C) 8 (D) 10

27. Find the value of a from the following system of equations

$$a + 1 = b$$
$$a = 2b + 3$$

(A) $a = 4$ (B) $a = -2$
(C) $a = 3$ (D) $a = -5$

28. Solve the following compound inequality: $6 > -X + \frac{5}{2} > -3$

(A) $\left[\frac{-7}{2}, \frac{11}{2}\right]$ (B) $\left(\frac{-7}{2}, \frac{11}{2}\right)$

(C)$\left(\frac{7}{2}, \frac{-11}{2}\right)$ (D)$\left(\frac{-7}{2}, \infty\right)$

29. Factor: $X^3 + 5X^2 - 4X - 20$
    (A)$(X+2)(X-2)(X+5)$ (B)$(X-2)^2(X+5)$
    (C)$(X+4)(X-5)X$ (D) None of the above

30. Find the perpendicular slope of the line $-5X - 6Y = -12$
    (A)$m = \frac{-5}{6}$ (B)$m = \frac{1}{5}$ (C)$m = \frac{6}{5}$
    (D)$m = 5$

Pert Test 4 answer key

1.D $\sqrt{3X-5} = 1$ $\left(\sqrt{3X-5}\right)^2 = (1)^2$ $3X - 5 = 1$ $3X = 1 + 5$

$3X = 6$ $X = \frac{6}{3} = 2$

2.A $2X^3 + 5X^2 - 8X - 20 = 0$ $X^2(2X+5) - 4(2X+5)$
$(X^2 - 4)(2X + 5) = 0$ You can continue factoring
$(X+2)(X-2)(2X+5) = 0$
$X + 2 = 0$ $X - 2 = 0$ $2X + 5 = 0$ $X = -2$
$X = 2$ $2X = -5$ $X = \frac{-5}{2}$

3. B $F(X) = -3X^2 + X - 5$ $F(a-2) = -3(a-2)^2 + (a-2) - 5$

$F(a-2) = -3(a-2)(a-2) + (a-2) - 5$
$= -3(a^2 - 2a - 2a + 4) + a - 2 - 5$
$= -3(a^2 - 4a + 4) + a - 2 - 5$
$= -3a^2 + 12a - 12 + a - 7$
$= -3a^2 + 12a + a - 12 - 7$

$$= -3a^2 + 13a - 19$$

4. A $d = \sqrt{(Y_2 - Y_1)^2 + (X_2 - X_1)^2)}$ $(-3,5)\ (2,6)$

$$d = \sqrt{(6-5)^2 + \left(2-(-3)\right)^2}$$
$$d = \sqrt{(1)^2 + (2+3)^2}$$
$$d = \sqrt{(1)^2 + (5)^2} = \sqrt{1+25} = \sqrt{26}$$

5. A $F(X) = \frac{3}{X+5}$ Remember the domain is defined as the values of X restricted in a function, in this case the values that make the denominator equal to zero.

$$X + 5 \neq 0$$
$$X \neq -5$$
$$(-\infty, -5) \cup (-5, \infty)$$

6. B
$$3X - Y = -2$$
$$-X + 5Y = 24$$

Multiply the second equation by 3

$$3X - Y = -2$$
$$3(-X) + 3(5Y) = 3(24)$$

Add both equations to eliminate the X

$$3X - Y = -2$$
$$\underline{-3X + 15Y = 72}$$
$$14Y = 70$$

$$Y = \frac{70}{14} = 5$$

Since $Y = 5$ then substitute it into the first equation to find X

$$3X - 5 = -2 \qquad 3X = -2 + 5$$
$$3X = 3 \qquad X = \frac{3}{3} = 1$$

347

$$(1, 5)$$

7.A $-5 = 3X^2 - 8X$

Move all terms to one side

$$3X^2 - 8X + 5 = 0$$

Multiply 3 by 5 = 15 then the factors of 15 that add up to -8 are -3 and -5.

$$3X^2 - 3X - 5X + 5 = 0$$
$$3X(X - 1) - 5(X - 1) = 0$$
$$(3X - 5)(X - 1) = 0$$
$$3X - 5 = 0$$
$$3X = 5$$
$$X = \frac{5}{3}$$
$$X - 1 = 0$$
$$X = 1$$

8.D $Y \geq -3X + 5$

1. The first step is to graph both lines

$$X \geq -2$$

2. Both lines need to shaded to the right and above since the Inequalities are $\geq$ (greater than equal to)

3. Finally the intersection of both areas is the answer

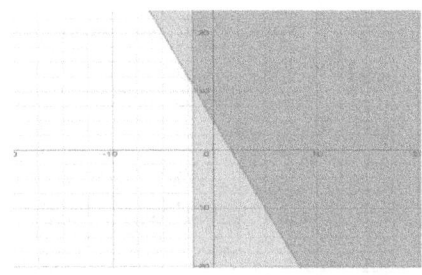

9. B $\dfrac{3}{X^2+3X-10} - \dfrac{X}{X^2+2X-15}$

1. Factor both denominators

$$\dfrac{3(X-3)}{(X+5)(X-2)(X-3)} - \dfrac{X(X-2)}{(X+5)(X-3)(X-2)}$$

2. Make both denominators equal to each other, don't forget whatever you do to the denominator you have to do it to the numerator

$$= \dfrac{3X-9-X^2+2X}{(X+5)(X-2)(X-3)} = \dfrac{-X^2+5X-9}{(X+5)(X-2)(X-3)}$$

10. C $\dfrac{8a+16}{4-a^2} = \dfrac{8(a+2)}{(2-a)(2+a)} = \dfrac{8}{2-a}$

11. D $Y = 2X^2 + 3X - 5$

$$2X^2 + 3X - 5 = 0$$

Factor by multiplying 2 and $-5 = -10$, the factors of -10 that add up to 3 are 5 and -2.

$$2X^2 + 3X - 5 = 0$$
$$X(2X+5) - 1(2X+5) = 0$$
$$(X-1)(2X+5) = 0$$
$$X - 1 = 0$$
$$\mathbf{X = 1}$$
$$2X + 5 = 0$$
$$2X = -5$$
$$X = \dfrac{-5}{2}$$

The intercepts are $(1, 0) \left(\dfrac{-5}{2}, 0\right)$

12. C $F(X) - G(X) = (3X - 2) - (-5X + 6)$
$$= 3X - 2 + 5X - 6$$
$$= 8X - 8$$

13. B   Make $X = 0$ to find Y $2Y - 3(0) = 36$

$2Y = 36$

$$Y = \frac{36}{2} = 18$$

$$(\mathbf{0, 18})$$

Make $Y = 0$ to find X $2(0) - 3X = 36$

$-3X = 36$ $X = \frac{36}{-3} = -12$

$$(\mathbf{-12, 0})$$

14. D $K_E = \frac{1}{2}MV^2$ $2(K_E) = 2\left(\frac{1}{2}\right)MV^2$ $2K_E = MV^2$

$\frac{2K_E}{M} = V^2$ $V = \sqrt{\frac{2K_E}{M}}$

15. A $\mathbf{3 + 2X = 3X - 5}$

16. A $\frac{a+b}{a-b} \div \frac{-a-b}{2a-2b}$

$\frac{a+b}{a-b} \times \frac{-2a-2b}{-a-b} = \frac{\cancel{a+b}}{\cancel{a-b}} \times \frac{2\cancel{(a-b)}}{-1\cancel{(a+b)}} = \frac{2}{-1} = \mathbf{-2}$

17. A $(8X - 3Y)^2 = (8X - 3Y)(8X - 3Y) =$
$64X^2 - 24XY - 24XY + 9Y^2 = \mathbf{64X^2 - 48XY + 9Y^2}$

18. A $(5X^2 - 2X + 3)(X - 1) =$
$5X^3 - 5X^2 - 2X^2 + 2X + 3X - 3$
$= \mathbf{5X^3 - 7X^2 + 5X - 3}$

19. D   The slope of line $x = -3$ is undefined, therefore the parallel line is also **undefined.**

20. D $\dfrac{X^4-256}{X^2-4X-32} = \dfrac{(X^2-16)(X^2+16)}{(X-8)(X+4)} = \dfrac{(X+4)(X-4)(X^2+16)}{(X-8)(X+4)} = \dfrac{(X-4)(X^2+16)}{X-8}$

21. A $(-1)^2(-1)^3(-1)^{-5} = (-1)^{2+3-(-5)} = (-1)^{5+5} = (-1)^{10} = $ **1**

22. D $\sqrt{729a^{17}b^{16}c^{11}}$ = rewriting the expression =
$\sqrt{(27)(27)a^{16} \cdot ab^{16} \cdot c^{10} \cdot c} = 27\sqrt{a^{16} \cdot a \cdot b^{16} \cdot c^{10} \cdot c}$
$27a^{\frac{16}{2}} \cdot b^{\frac{16}{2}} \cdot c^{\frac{10}{2}} \sqrt{ac} = \mathbf{27a^8 b^8 c^5 \sqrt{ac}}$

23. A $V_F^2 = V_I^2 + 2ad$ $\qquad V_F^2 - V_I^2 = 2ad \qquad d = \dfrac{V_F^2 - V_I^2}{2a}$

24. B $3X^2 + 6X - 1 = 0 \qquad$ Using the quadratic formula
$a = 3, b = 6, c = -1$
$X = \dfrac{-b \pm \sqrt{b^2 - 4ac}}{2a} = \dfrac{-6 \pm \sqrt{(6)^2 - 4(3)(-1)}}{2(3)}$
$= \dfrac{-6 \pm \sqrt{36 + 12}}{6} = \dfrac{-6 \pm \sqrt{48}}{6} = \dfrac{-6 \pm \sqrt{16(3)}}{6} = \dfrac{-6 \pm 4\sqrt{3}}{6}$
$X = \dfrac{-3 \pm 2\sqrt{3}}{3}$

25. A $X^2 + 3X + 4 = 0 \quad a = 1, b = 3, c = 4$
Find the discriminant $b^2 - 4ac = (3)^2 - 4(1)(4) = 9 - 16 = -7$
Since the discriminant is less than zero, **there are no real solutions.**

26. B $F(X) = 3X^3 + 6X^2 - 1$
$F(-1) = 3(-1)^3 + 6(-1)^2 - 1$
$F(-1) = 3(-1) + 6(1) - 1 = -3 + 6 - 1 = 3 - 1 = \mathbf{2}$

27. D $a + 1 = b$ Substituting for b in the second equation

$a = 2b + 3$ $a = 2(a+1) + 3,$ $a = 2a + 2 + 3$

$a = 2a + 5$ $a - 2a = 5$

$-a = 5$ $\boldsymbol{a = -5}$

28. B $6 > -X + \frac{5}{2} > -3$

Multiply every term by 2

$12 > -2X + 5 > -6$

$12 - 5 > -2X > -6 - 5$

$7 > -2X - 11$

$\frac{7}{-2} < X < \frac{-11}{-2}$

(The sign flips, since you are dividing by a negative number)

$\frac{-7}{2} < X < \frac{11}{2}$

$\left(\frac{-7}{2}, \frac{11}{2}\right)$

29. A   Factor: $X^3 + 5X^2 - 4X - 20$

By using the grouping method, you can factor $X^3 + 5X^2$ with the greatest common factor X^2, then you can factor $-4X - 20$ with the greatest common factor -4.

$X^3 + 5X^2 - 4X - 20 = X^2(x + 5) - 4(x + 5) = (X^2 - 4)(x + 5)$

Then you can factor $X^2 - 4$ as a difference of squares (x+2)(x-2)

$= (\boldsymbol{x + 2})(\boldsymbol{x - 2})(\boldsymbol{x + 5})$

30. C   Find the perpendicular slope of the line $-5X - 6Y = -12$

The first step is to solve for Y by isolating the Y term.

$-6Y = 5X - 12$

$$\frac{-6}{-6}Y = \frac{5}{-6}X - \frac{12}{-6}$$

$Y = \frac{-5}{6}X + \frac{12}{6}$ The slope is m= $\frac{-5}{6}$, then the perpendicular slope is the negative reciprocal.

$$m \perp = \frac{6}{5}$$

Practice Test 5

1. Evaluate: $-3^n + 2^{-n}$ If $n = -1$

    (A) $\frac{3}{5}$ (B) $\frac{5}{3}$ (C) $\frac{1}{3}$ (D) 0

2. Solve for X: $2X + \frac{1}{5}X = 6 - \frac{X}{3}$

    (A) $X = \frac{5}{76}$ (B) $X = \frac{6}{71}$ (C) $X = \frac{45}{19}$ (D) $X = 3$

3. Factor $X^4 - Y^4$

    (A) $(X^2 + Y^2)(X^2 + Y^2)$ (B) $(X + Y)(X + Y)(X^2 - Y^2)$
    (C) $(X - Y)^4$ (D) $(X + Y)(X - Y)(X^2 + Y^2)$

4. Build the equation of the line that passes through $(0, -3)$ and $(0, -5)$

    (A) $Y = \frac{3}{5}$ (B) $Y = 0$ (C) $X = 0$

    (D) $X = \frac{-3}{5}$

5. Simplify: $\frac{2X^3 - 2X^2 - 12X}{X^3 - 9X}$

    (A) $\frac{2(X+2)}{X+3}$ (B) $\frac{X+2}{X+3}$ (C) $\frac{X-2}{X-3}$ (D) $\frac{X}{3}$

6. Solve for b_2 $A = \left(\frac{b_1 + b_2}{2}\right)h$

    (A) $b_2 = \frac{2A - b_1}{h}$ (B) $b_2 = \frac{2+A}{b}$ (C) $b_2 = \frac{A}{b} - b_1$ (D) $b_2 = 2A/h - b_1$

7. Solve the following system of equations:

    $3X - 2Y = 10$

$$6X = 4Y + 30$$

(A) (2,3) (B) No Solution (C) (3,2) (D) Infinite Solutions

8. Simplify $(3a^{-5}b^{-2}c^5)^{-2}(3a^4b^2c^7)^{-1}$

   (A) $\dfrac{a^4b^2}{ac^{17}}$ (B) $\dfrac{a^6b^2}{27c^{17}}$ (C) $\dfrac{a^4b}{27c^{10}}$ (D) $\dfrac{27a^4b^2}{c^{17}}$

9. Graph the line using the intercepts: $-2X + Y = -20$

   (A)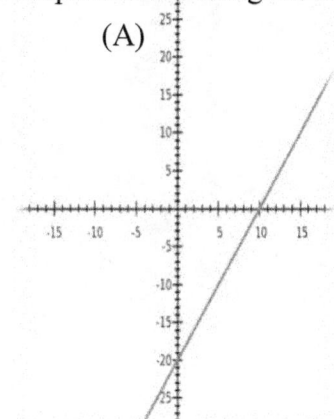

   (B)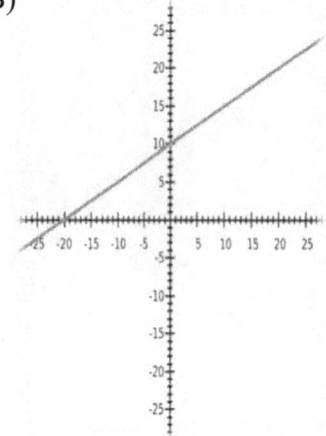

   (C)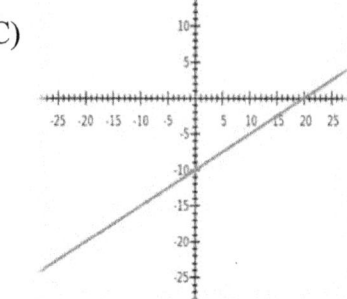

   (D) None of the above.

10. Solve: $3X - 12 = 6(-2X + 6) + 5$

    (A) $X = 0$ (B) $X = \dfrac{53}{15}$ (C) $X = -\dfrac{5}{6}$

    (D) $X = 5$

11. Simplify: $\dfrac{3}{X+2} - \dfrac{4}{X^2-4}$

(A) $\dfrac{3X+10}{X^2-4}$ (B) $\dfrac{3X-10}{X+2}$ (C) $\dfrac{3X-10}{X^2-4}$ (D) $\dfrac{3X+2}{X^2-4}$

12. Graph: $2Y + 3X \leq 10$

A)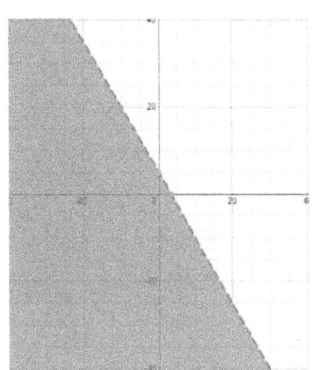

B)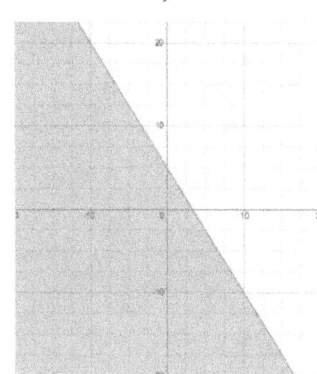

C)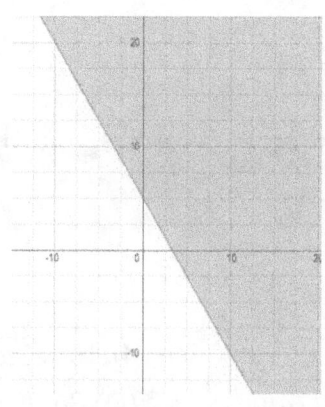

D)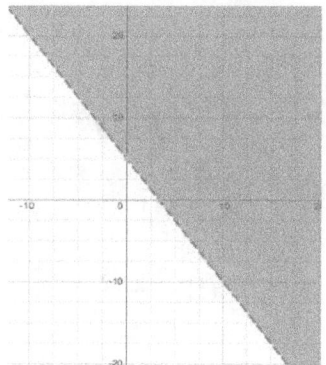

13. Multiply: $(X^2Y - Y^3)(X^2Y + Y^3)$

(A) $X^2Y - Y^6$ (B) $X^2Y - Y^9$ (C) $X^4Y^2 - Y^6$ (D) $X^4Y^2 + Y^9$

14. Factor: $16X^3Y^4 - 14X^2Y^2 - 2XY$

(A) $2XY(8X^2Y^3 - 7XY - 1$ (B) $2XY(8X^2Y^3 + 7XY + 1)$

(C) $XY(8X^2Y - 7X - XY)$ (D) None of the above.

15. Solve: $-3|X + 6| = -6$

(A) No Solution (B) $\{-8, -4\}$ (C) $\{-8, 4\}$ (D) $\{-4\}$

16. Solve for X: $8X^2 + 37X = 15$

    (A) $\left\{-5, \dfrac{3}{8}\right\}$ (B) $\left\{-5, \dfrac{-3}{8}\right\}$ (C) $\{-5\}$ (D) No real Solution

17. Simplify: $\sqrt{32a^7b^{11}}$

    (A) $16a^6b^5\sqrt{2ab}$ (B) $4a^3b^5\sqrt{2ab}$
    (C) $4a^5b^3\sqrt{2ab}$ (D) $8a^3b^5\sqrt{2ab}$

18. Simplify: $\sqrt{147} + \sqrt{192} - \sqrt{48}$

    (A) $-11\sqrt{3}$ (B) $11\sqrt{3}$ (C) $19\sqrt{3}$ (D) $4\sqrt{3}$

19. Rationalize the denominator: $\dfrac{3}{\sqrt{5}-1}$

    (A) $\dfrac{3+3\sqrt{5}}{4}$ (B) $\dfrac{3-3\sqrt{5}}{4}$

    (C) $3 + 3\sqrt{5}$ (D) $\dfrac{-3\sqrt{5}}{4}$

20. Simplify: $\left(\dfrac{3b^{-10}c^8}{4a^{-6}b^{-3}c^{-2}}\right)^{-2}$

    (A) $\dfrac{16}{9a^4b^{14}c^4}$ (B) $\dfrac{3b^{14}}{4a^{12}c^{20}}$

    (C) $\dfrac{9b^{14}}{16a^{12}c^4}$ (D) $\dfrac{16b^{14}}{9a^{12}c^{20}}$

21. Solve: $-3\sqrt{X+5} = -15$

    (A) $X = 20$ (B) $X = 10$
    (C) $X = -5$ (D) No Solution

22. Solve: $\dfrac{3}{X+3} - 5 = \dfrac{-2}{X+3}$

(A) $X = 2$ (B) $X = -2$
(C) $X = 0$ (D) No solution

23. If $F(X) = 2X^2 + 7X - 15$ and $G(X) = X + 5$. Find $\frac{F(X)}{G(X)}$

(A) X (B) $\frac{X-3}{X+5}$ (C) $X + 5$ (D) $2X - 3$

24. Solve: $-3|X + 5| = -6$
(A) $\{-7, -3\}$ (B) $\{-7\}$ (C) $\{3, 7\}$ (D) No Solution.

25. Simplify: $\frac{3X^2+X-2}{X^2-1} \div \frac{3X-2}{2X-2}$

(A) 2 (B) -2 (C) $\frac{2}{X+1}$ (D) $\frac{3}{X-2}$

26. Solve for the values of X: $-16X^2 = 30X - 25$

(A) $\left\{\frac{-5}{2}, \frac{5}{8}\right\}$ (B) $\left\{\frac{-5}{8}, \frac{5}{2}\right\}$ (C) $\left\{\frac{5}{8}, 3\right\}$ (D) $\left\{\frac{-5}{2}, 0\right\}$

27. Find the vertex of the following function f(x)= $-2X^2 + 12X+5$
(A) (3,-49) (B) (-3,-49)
(C) (23,3) (D) (3,23)

28. A cupcake store has the following revenue function
R(x) = $-3X^2 + 12X+10$, where x is the amount of cupcakes sold every day. Find the Maximum Revenue for the store. (Hint: Find the Maximum point)

(A) $2 (B) $22 (C) $13 (D) $300

29. Find g(-1) if $g(x) = -3^{2x}$

(A) $\frac{-1}{9}$ (B) $\frac{1}{9}$ (C) 9 (D) -9

30. Solve the following system of equations:
$$5X - 8Y = 2$$
$$2X = 3Y + 1$$

(A) (2,2) (B) (-1,-2)
(C) (1,2) (D) (2,1)

Practice Test 5 answer key

1. B $\quad -3^{(-1)} + 2^{-(-1)} = -3^{-1} + 2^1 = \frac{-1}{3} + 2 = \frac{-1}{3} + \frac{2(3)}{3} = \frac{-1}{3} + \frac{6}{3} = \frac{5}{3}$

2. C $\quad 2X + \frac{1}{5}X = 6 - \frac{X}{3}$

   Multiply all terms by 15

   $$15(2X) + 15\left(\frac{X}{5}\right) = 15(6) - 15\left(\frac{X}{3}\right)$$

   $$30X + 3X = 90 - 5X$$

   $$33X = 90 - 5X$$

   $$33X + 5X = 90$$

   $$38X = 90$$

   $$X = \frac{90}{38} = \frac{45}{19}$$

3. D Factor $X^4 - Y^4$

   $$(X^2 - Y^2)(X^2 + Y^2)$$

   $$(X + Y)(X - Y)(X^2 + Y^2)$$

4. C Find the slope $m = \frac{Y_2 - Y_1}{X_2 - X_1}$

   $$m = \frac{-5-(-3)}{0-0} = \frac{-5+3}{0} = \frac{-2}{0} = \text{Undefined}$$

   Since the line is undefined, the line is vertical, using any point either $(0, -3)$ or $(0, -5)$, the

   Answer is $X = 0$

5. A Factor $\frac{2X^3 - 2X^2 - 12X}{X^3 - 9X}$

$$\frac{2X^3-2X^2-12X}{X^3-9X} = \frac{2X(X^2-X-6)}{X(X^2-9)} = \frac{2X(X-3)(X+2)}{X(X+3)(X-3)} = \frac{2(X+2)}{(X+3)}$$

6. D Solve for b_2 $A = \left(\frac{b_1+b_2}{2}\right) \cdot h$

$$A = \left(\frac{b_1 + b_2}{2}\right) \cdot h$$

$$2A = (b_1 + b_2) \cdot h$$

$$\frac{2A}{h} = b_1 + b_2$$

$$b_2 = \frac{2A}{h} - b_1$$

7. B

Multiply the top equation by -2

$$3X - 2Y = 10$$
$$6X = 4Y + 30$$

to eliminate the variable "X"

$$-2(3X) - 2(-2Y) = -2(10)$$

$$-6X + 4Y = -20$$
$$6X - 4Y = 30$$

$$0 \neq 10$$

Since the variables X and Y disappear and 0 is not equal to 10, the system has no solutions.

8. B Simplify $(3a^{-5}b^{-2}c^5)^{-2}(3a^4b^2c^7)^{-1}$

$(3^{-2} \cdot a^{10} \cdot b^4 \cdot c^{-10})(3^{-1} \cdot a^{-4} \cdot b^{-2} \cdot c^{-7}) =$

$3^{-2-1} \cdot a^{10-4} \cdot b^{4-2} \cdot c^{-10-7}$

$$3^{-3} \cdot a^6 \cdot b^2 \cdot c^{-17} = \frac{1 \cdot a^6 \cdot b^2}{3^3 c^{17}} = \frac{1 \cdot a^6 \cdot b^2}{27 c^{17}} = \frac{a^6 \cdot b^2}{27 c^{17}}$$

9. A $\quad -2X + Y = -20$

In order to find the "X" intercept make $Y = 0$

$$-2X + (0) = -20$$
$$-2X = -20$$
$$X = 10$$
$$(\mathbf{10}, \mathbf{0})$$

In order to find the "Y" intercept make $X = 0$

$$-2(0) + Y = -20$$
$$0 + Y = -20$$
$$Y = -20$$
$$(\mathbf{0}, -\mathbf{20})$$

Finally, you can graph

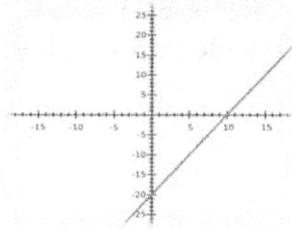

10. B
$$3X - 12 = 6(-2X + 6) + 5$$
$$3X - 12 = -12X + 36 + 15$$
$$3X + 12X = 36 + 15 + 12$$
$$15X = 53$$
$$X = \frac{53}{15}$$

11. C Simplify: $\quad \frac{3}{X+2} - \frac{4}{X^2-4}$

$\frac{3}{X+2} - \frac{4}{X^2-4} = \quad$ Making the denominators the same

$$\frac{3(X-2)}{(X+2)(X-2)} - \frac{4}{(X+2)(X-2)} = \frac{3X-6-4}{(X+2)(X-2)}$$

$$\frac{3X-10}{(X+2)(X-2)} = \frac{3X-10}{X^2-4}$$

12. B   Graph: $2Y + 3X \leq 10$

Solve for Y

$$2Y \leq -3X + 10$$

$$Y \leq \frac{-3X}{2} + \frac{10}{2}$$

$$Y \leq \frac{-3X}{2} + 5$$

Since the inequality is "less than or equal to" then the line is **solid** and you need to **shade below** the line.

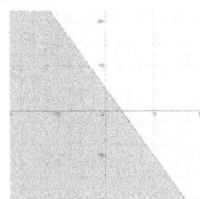

13. C

Multiply:
$$(X^2Y - Y^3)(X^2Y + Y^3) =$$
$$(X^2Y)(X^2Y) + X^2Y \cdot Y^3 - X^2Y \cdot Y^3 - Y^3Y^3$$
$$= X^2Y^2 + X^2Y^4 - X^2Y^4 - Y^6$$
$$= X^4Y^2 - Y^6$$

14. A   Factor: $16X^3Y^4 - 14X^2Y^2 - 2XY$

The G.C.F (Greatest Common Factor) is $2XY$

$$2XY(8X^2Y^3 - 7XY - 1)$$

15. B

Solve: $-3|X + 6| = -6$

$$|X+6| = \frac{-6}{-3} \qquad |X+6| = 2$$

$$X + 6 = 2 \quad \text{and} \quad X + 6 = -2$$
$$X = 2 - 6 \qquad\qquad X = -2 - 6$$
$$\mathbf{X = -4} \qquad \mathbf{X = -8}$$

16. A   Solve for X: $8X^2 + 37X = 15$

$$8X^2 + 37X - 15 = 0$$

1) Multiply 8 and $-15 = (8)(-15) = -120$

2) Find the factors of -120 that will add up to 37, in this case 40 and -3, since $(40)(-3) = -120$ and $40 - 3 = 37$.

3) Rewrite $37X$ as $40X - 3X$ and factor.

$$8X^2 + 40X - 3X - 15 = 0$$
$$8X(X + 5) - 3(X + 5) = 0$$
$$(8X - 3)(X + 5) = 0$$

Finally $8X - 3 = 0$ and $X + 5 = 0$

$$X = \frac{3}{8} \qquad \text{and} \qquad X = -5$$

$$\left\{-5, \frac{3}{8}\right\}$$

17. B   Simplify: $\sqrt{32a^7b^{11}} = \sqrt{16} \cdot \sqrt{2} \sqrt{a^6} \cdot \sqrt{a} \cdot \sqrt{b^{10}} \cdot \sqrt{b}$
$$= \mathbf{4a^3 b^5 \sqrt{2ab}}$$

18. B $\sqrt{147} + \sqrt{192} - \sqrt{48} = \sqrt{49}\sqrt{3} + \sqrt{64}\sqrt{3} - \sqrt{16}\sqrt{3} =$
$7\sqrt{3} + 8\sqrt{3} - 4\sqrt{3} = 15\sqrt{3} - 4\sqrt{3} = \mathbf{11\sqrt{3}}$

19. A   Rationalize the denominator: $\frac{3}{\sqrt{5}-1}$

$\frac{3}{\sqrt{5}-1}$, to rationalize the denominator multiply the denominator and numerator by the conjugate.

The conjugate is the same expression with a different sign. In this case: $\sqrt{5} + 1$

$$\frac{3}{\sqrt{5}-1} \frac{(\sqrt{5}+1)}{(\sqrt{5}+1)} = \frac{3\sqrt{5}+3}{\sqrt{25}-\sqrt{5}+\sqrt{5}-1} = \frac{3\sqrt{5}+3}{5-1} = \frac{3\sqrt{5}+3}{4}$$

20. D Simplify: $\left(\frac{3b^{-10}c^8}{4a^{-6}b^{-3}c^{-2}}\right)^{-2}$ = Distribute the -2 to all the exponents (including the numbers 3 and 4)

$\left(\frac{3^{-2}b^{20}c^{-16}}{4^{-2}a^{12}b^6c^4}\right) = \frac{3^{-2}b^{20}c^{-16}}{4^{-2}a^{12}b^6c^4}$ = Switch from bottom to top and vice-versa the negative exponent to make them positive

$$\frac{4^2 b^{20}}{3^2 a^{12} b^6 c^4 c^{16}} = \frac{16 b^{14}}{9 a^{12} c^{20}}$$

21. A Solve: $-3\sqrt{X+5} = -15$ Divide -3 on both sides

$\frac{-3}{-3}\sqrt{X+5} = \frac{-15}{-3} = \sqrt{X+5} = 5$

Finally square both sides to eliminate the square root
$X + 5 = 25 =$ X=25-5= **20**

22. B Solve: $\frac{3}{X+3} - 5 = \frac{-2}{X+3}$

Multiply every term by the Least common denominator x+3

$(x+3)\frac{3}{X+3} - 5(x+3) = \frac{-2}{X+3}(x+3)$

$3 - 5(x+3) = -2$ $3 - 5x - 15 = -2$ $-5x = -2-3+15$

$-5x = 10$ $x = -2$

23) D If $F(X) = 2X^2 + 7X - 15$ and $G(X) = X + 5$. Find $\frac{F(X)}{G(X)}$

$\frac{F(X)}{G(X)} = \frac{F(X)}{G(X)} = \frac{2x^2+10x-3x-15}{x+5} = \frac{2x(x+5)-3(x+5)}{x+5} = \frac{(2x-3)(x+5)}{x+5} =$ **2x-3**

24) Solve: $-3|X+5| = -6$

Divide both sides by -3, $|X+5| = 2$, now you can solve both equations

X+5=2 and X+5=-2

X= -3 and X= -7

25)A Simplify $\frac{3X^2+X-2}{X^2-1} \div \frac{3X-2}{2X-2}$

$\frac{3X^2+X-2}{X^2-1} \cdot \frac{2X-2}{3X-2}$. The first step is to flip the fraction and then multiply. You need to factor each term.

$3X^2 + X - 2 = 3X^2 + 3X - 2x - 2 = 3x(x+1) - 2(x+1) = (3x-2)(x+1)$

$\frac{(3x-2)(x+1)}{(x+1)(x-1)} \cdot \frac{2(X-1)}{3X-2} = \frac{(3x-2)(x+1)}{(x+1)(x-1)} \cdot \frac{2(X-1)}{3X-2} = 2$

26.A Solve for the values of $X: -16X^2 = 30X - 25$

The first step is to move all the $16X^2 + 30X - 25 = 0$, then you can either factor or use the quadratic formula

Let's practice the quadratic formula:

$a = 16, b = 30, c = -25$

$\frac{-b \pm \sqrt{b^2-4ac}}{2a} = \frac{-30 \pm \sqrt{(30)^2-4(16)(-25)}}{2(16)} = \frac{-30 \mp \sqrt{2500}}{32} = \frac{-30 \mp \sqrt{2500}}{32} = \frac{-30 \mp 50}{32} = $ -80/32

and 20/32 , your answers are $\left\{\frac{-5}{2}, \frac{5}{8}\right\}$

27.D Find the vertex of the following function F(x)= $-2X^2 + 12X+5$

The formula for vertex to find x is $-b/2a$ since b= 12 and a=-2.

Then x=(-12/(2(-2))= 12/4= 3.

To find the value of y, then plug 3 into the original formula

$F(3) = -2(3)^2 + 12(3)+5 = 23$

The vertex is (3,23)

28.B A cupcake store has the following revenue function

R(x) = $-3X^2 + 12X+10$, where x is the amount of cupcakes sold every day. Find the Maximum Revenue for the store. (Hint: Find the Maximum point) The maximum point is found with the vertex formula. Let's find the value of X first.

-b/2a= -12/(2(-3))= -12/-6= 2

Then find R(2) = -3(4) + 12(2) + 10 = -12+24+10=$ **22**

29.A Find G(-1) if $G(x) = -3^{2x}$

$G(-1) = -3^{2(-1)} = -3^{(-2)} = \frac{1}{-3^{(2)}} = \frac{-1}{9}$

30.D Solve the following system of equations:

$$5X - 8Y = 2$$
$$2X = 3Y + 1$$

To solve the following system you need to use either elimination or substitution.

Let's practice substitution.

You can use X= 3/2Y + ½ and then plug it into the other equation 5x-8y= 2

$$5(3/2Y + ½) - 8Y = 2$$

15/2Y + 5/2-8Y= 2, then multiply all the terms by 2 to eliminate the fractions

15Y+ 5-16Y = 4, finally just solve for Y ,-Y+5= 4, -Y=-1, Y=1.

Since you have Y=1, then X= 3(1)/2 + ½= 3/2 + ½= 4/2 = 2

Final answer **(2,1)**

Cheat Sheets

Properties of Logarithms:
1. $\log_a(MN) = \log_a M + \log_a N$
2. $\log_a\left(\frac{M}{N}\right) = \log_a M - \log_a N$
3. $\log_a M^r = r \log_a M$
4. $\log_a M = \frac{\log M}{\log a} = \frac{\ln M}{\ln a}$

Arithmetic Sequence:

$$a + (a+d) + (a+2d) + \cdots + [a + (n-1)d] = na + \frac{n(n-1)}{2}d$$

Geometric Sequence:

$$a + ar + ar^2 + \cdots + ar^{n-1} = a\frac{1-r^n}{1-r}$$

Geometric Series: If $|r| < 1$

$$a + ar + ar^2 + \cdots = \sum_{k=1}^{\infty} ar^{k-1} = \frac{a}{1-r}$$

Permutations/Combinations:

$0! = 1 \qquad 1! = 1$

$n! = n(n-1) \cdot \ldots \cdot (3)(2)(1)$

$P(n,r) = \dfrac{n!}{(n-r)!}$

$C(n,r) = \binom{n}{r} = \dfrac{n!}{(n-r)!\,r!}$

Trigonometric Functions:

Of an acute angle:

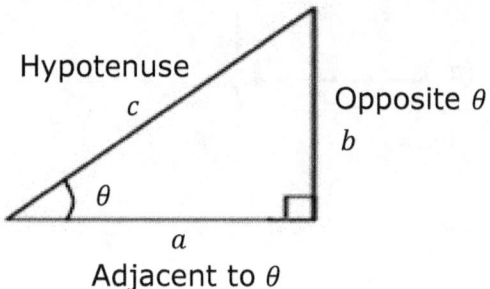

$$\sin \theta = \frac{b}{c} = \frac{Opposite}{Hypotenuse} \qquad \cos \theta = \frac{a}{c} = \frac{Adjacent}{Hypotenuse}$$

$$\tan \theta = \frac{b}{a} = \frac{Opposite}{Adjacent} \qquad \csc \theta = \frac{c}{b} = \frac{Hypotenuse}{Opposite}$$

$$\sec \theta = \frac{c}{a} = \frac{Hypotenuse}{Adjacent} \qquad \cot \theta = \frac{a}{b} = \frac{Adjacent}{Opposite}$$

Of a general angle:

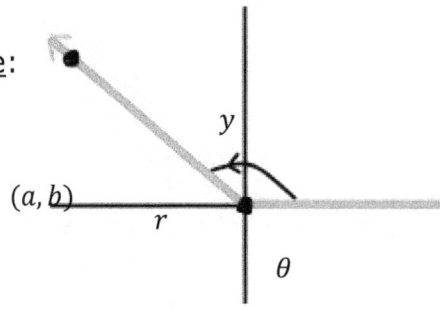

$\sin\theta = \dfrac{b}{r}$ $\cos\theta = \dfrac{a}{r}$ $\tan\theta = \dfrac{b}{a}, a \neq 0$

$\csc\theta = \dfrac{r}{b}, a \neq 0$ $\sec\theta = \dfrac{r}{a}, a \neq 0$ $\cot\theta = \dfrac{a}{b}, a \neq 0$

Trigonometric Identities:

Fundamental Identities:

$\tan\theta = \dfrac{\sin\theta}{\cos\theta}$ $\cot\theta = \dfrac{\cos\theta}{\sin\theta}$ $\csc\theta = \dfrac{1}{\sin\theta}$

$\sec\theta = \dfrac{1}{\cos\theta}$ $\cot\theta = \dfrac{1}{\tan\theta}$

$\sin^2\theta + \cos^2\theta = 1$

$\tan^2\theta + 1 = \sec^2\theta$

$\cot^2\theta + 1 = \csc^2\theta$

Half-Angle Formulas:

$$\sin\frac{\theta}{2} = \pm\sqrt{\frac{1-\cos\theta}{2}}$$

$$\cos\frac{\theta}{2} = \pm\sqrt{\frac{1+\cos\theta}{2}}$$

$$\tan\frac{\theta}{2} = \frac{1-\cos\theta}{\sin\theta}$$

Double-Angle Formulas:

$\sin(2\theta) = 2\sin\theta\cos\theta$

$\cos(2\theta) = \cos^2\theta - \sin^2\theta$

$\cos(2\theta) = 2\cos^2\theta - 1$

$\cos(2\theta) = 1 - 2\sin^2\theta$

$\tan(2\theta) = \dfrac{2\tan\theta}{1-\tan^2\theta}$

Even-Odd Identities:

$\sin(-\theta) = -\sin\theta$ $\csc(-\theta) = -\csc\theta$

$\cos(-\theta) = \cos\theta$ $\sec(-\theta) = \sec\theta$

$\tan(-\theta) = -\tan\theta$ $\cot(-\theta) = -\cot\theta$

Product-to-Sum Formulas:

$\sin\alpha\sin\beta = \frac{1}{2}[\cos(\alpha-\beta) - \cos(\alpha+\beta)]$

$\cos\alpha\cos\beta = \frac{1}{2}[\cos(\alpha-\beta) + \cos(\alpha+\beta)]$

$\sin\alpha\sin\beta = \frac{1}{2}[\sin(\alpha+\beta) + \sin(\alpha-\beta)]$

Sum and Difference Formulas:

$\sin(\alpha + \beta) = \sin\alpha \cos\beta + \cos\alpha \sin\beta$

$\sin(\alpha - \beta) = \sin\alpha \cos\beta - \cos\alpha \sin\beta$

$\cos(\alpha + \beta) = \cos\alpha \cos\beta - \sin\alpha \sin\beta$

$\cos(\alpha - \beta) = \cos\alpha \cos\beta + \sin\alpha \sin\beta$

$\tan(\alpha + \beta) = \frac{\tan\alpha + \tan\beta}{1 - \tan\alpha \tan\beta}$

$\tan(\alpha - \beta) = \frac{\tan\alpha - \tan\beta}{1 + \tan\alpha \tan\beta}$

Sum-to-Product Formulas:

$\sin\alpha + \sin\beta = 2 \sin\frac{\alpha+\beta}{2} \cos\frac{\alpha-\beta}{2}$

$\sin\alpha - \sin\beta = 2 \sin\frac{\alpha-\beta}{2} \cos\frac{\alpha-\beta}{2}$

$\cos\alpha + \cos\beta = 2 \cos\frac{\alpha+\beta}{2} \cos\frac{\alpha-\beta}{2}$

$\cos\alpha - \cos\beta = -2 \sin\frac{\alpha+\beta}{2} \sin\frac{\alpha-\beta}{2}$

Solving Triangles:

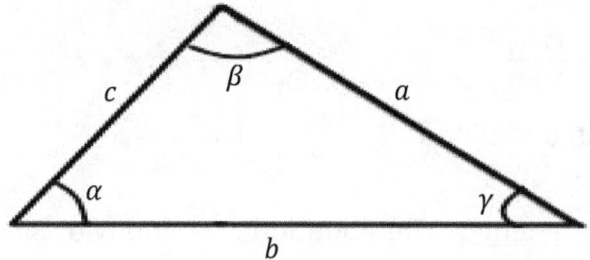

Law of Sines: $$\frac{\sin \alpha}{a} = \frac{\sin \beta}{b} = \frac{\sin \gamma}{c}$$

Law of Cosines: $$c^2 = a^2 + b^2 - 2ab \cdot \cos \gamma$$

Formulas/Equations:

Distance Formula: If $P_1 = (x_1, y_1)$ and $P_2 = (x_2, y_2)$, the distance from P_1 to P_2 is

$$d(P_1, P_2) = \sqrt{(x_2 - x_1)^2 + (y_2 - y_1)^2}$$

Standard equation of a Circle: The standard of a circle of radius r with center (h, k) is

$$(x - h)^2 + (y - k)^2 = r^2$$

Slope Formula: The slope m of the line containing the points $P_1 = (x_1, y_1)$ and $P_2 = (x_2, y_2)$ is

$$m = \frac{y_2 - y_1}{x_2 - x_1} \qquad \text{if } x_1 \neq x_2$$

$$m \text{ is undefined} \qquad \text{if } x_1 = x_2$$

Point-Slope equation of a Line: The equation of a line with slope m containing the point (x_1, y_1) is

$$y - y_1 - m(x - x_1)$$

Slope-Intercept equation of a Line: The equation of a line with slope m and y-intercept b is

$$y = mx + b$$

Quadratic Formula:

The solution(s) of the quadratic equation $ax^2 + bx + c = 0$, where $a \neq 0$ are

$$x = \frac{-b \pm \sqrt{b^2 - 4ac}}{2a}$$

If $b^2 - 4ac > 0$, there are two unequal real solutions.

If $b^2 - 4ac = 0$, there is a repeated real solution.

If $b^2 - 4ac < 0$, there are two complex solutions that are not real.

Geometry Formulas:

Circle: 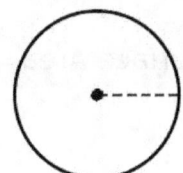 r = Radius, A = Area, C = Circumference

$A = \pi r^2$, $C = 2\pi r$

Triangle: 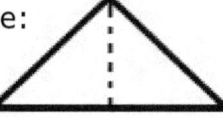 b = Base, h = Altitude(Height), A = Area

$A = \frac{1}{2}bh$

Rectangle:  l = Length, w = Width, A = Area

P = Perimeter

$A = lw$ $P = 2l + 2w$

Rectangular Box: 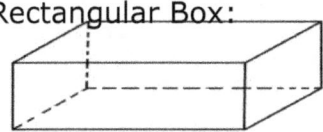 l = Length, w = Width, h = Height

V = Volume, S = Surface area

$V = lwh$ $S = 2lw + 2lh + 2wh$

Sphere:  r = Radius, V = Volume, S = Surface Area

$V = \frac{4}{3}\pi r^3$ $S = 4\pi r^2$

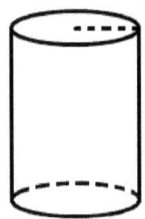

r =Radius, V =Volume, S =Surface Area

h =Height

$V = \pi r^2 h \qquad S = 2\pi r^2 + 2\pi rh$

Library of Functions:

Identity Function: $f(x) = x$

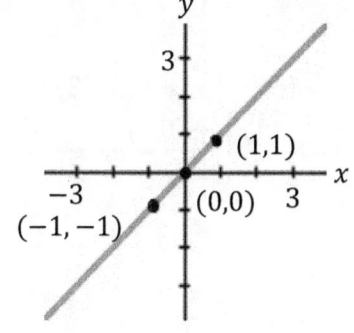

Square Function: $f(x) = x^2$

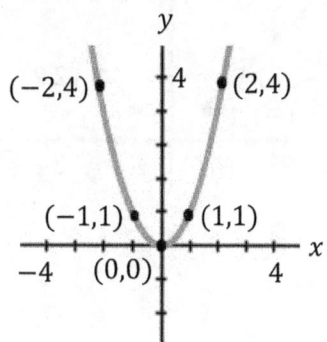

Cube Function: $f(x) = x^3$

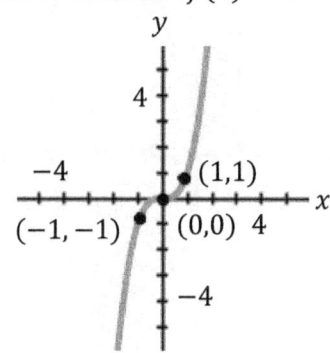

Square Root Function: $f(x) = \sqrt{x}$

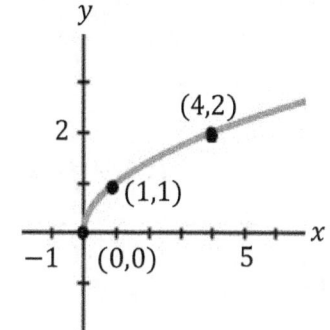

Reciprocal Function: $f(x) = \frac{1}{x}$

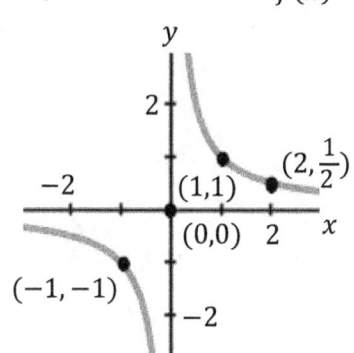

Cube Root Function: $f(x) = \sqrt[3]{x}$

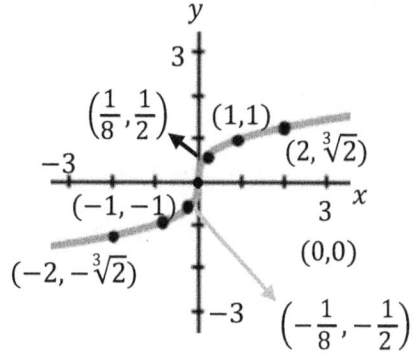

Absolute Value Function: $f(x) = |x|$

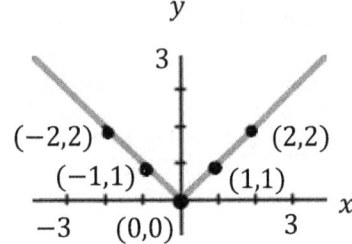

Exponential Function: $f(x) = e^x$

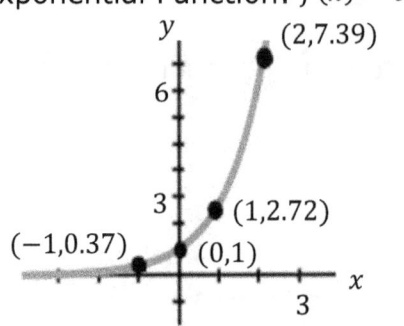

Natural Logarithm Function: $f(x) = \ln x$

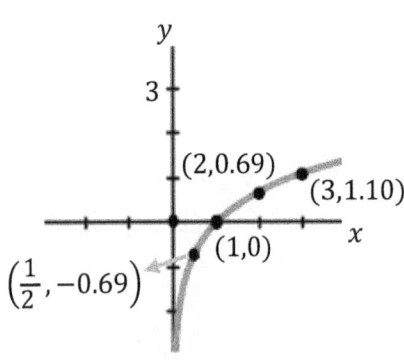

Sine Function: $f(x) = \sin x$

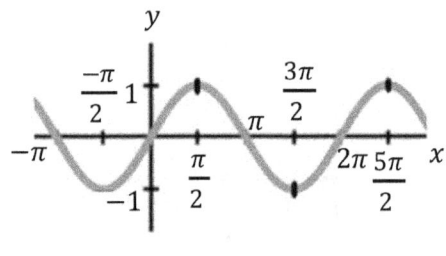

Cosine Function: $f(x) = \cos x$

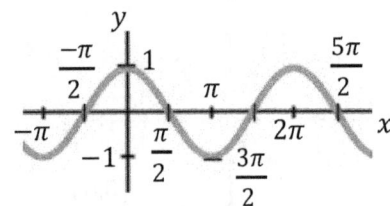

Tangent Function: $f(x) = \tan x$

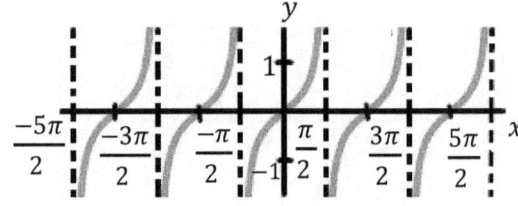

Cosecant Function: $f(x) = \csc x$

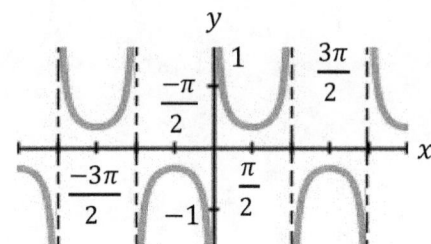

Secant Function: $f(x) = \sec x$

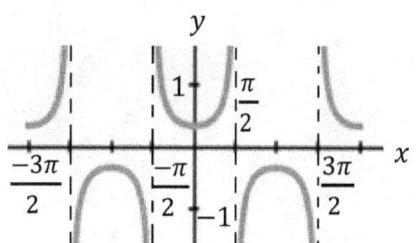

Cotangent Function: $f(x) = \cot x$

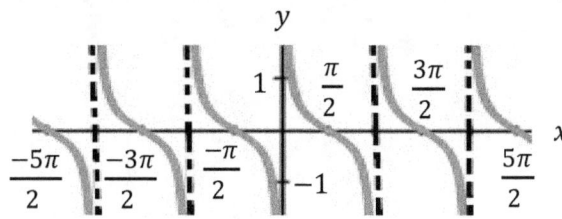

www.ingramcontent.com/pod-product-compliance
Lightning Source LLC
Chambersburg PA
CBHW080649190526
45169CB00006B/2043